A D.
Western Alchemy

A Dictionary of Western Alchemy

JORDAN STRATFORD

Foreword by JEFFREY S. KUPPERMAN, Ph.D.

Theosophical Publishing House
Wheaton, Illinois * Chennai, India

Copyright © 2011 by Jordan Stratford

First Quest Edition 2011

All rights reserved. No part of this book may be reproduced in any manner without written permission except for quotations embodied in critical articles or reviews. For additional information write to

Quest Books
Theosophical Publishing House
P. O. Box 270
Wheaton, IL 60187-0270

www.questbooks.net

Cover image: *Splendor Solis*, folio 10, © The British Library Board, Harley 3469f10. Illustrations in this book detail sections of *The Ripley Scroll of the Royal College of Physicians of Edinburgh*, c. 1640.
Cover design by Kirsten Hansen Pott
Typesetting by Wordstop Technologies, Chennai, India

Library of Congress Cataloging-in-Publication Data

Stratford, Jordan.
A dictionary of Western alchemy / Jordan Stratford; foreword by Jeffrey S. Kupperman.–1st Quest ed.
  p. cm.
Includes bibliographical references.
ISBN 978-0-8356-0897-8
1. Alchemy–Dictionaries.   I. Title.
QD23.5.S77 2011
540.1'12--dc23   2011019866

5 4 3 2 1 * 11 12 13 14 15

Printed in the United States of America

Dedication

To the Brethren of Victoria Columbia Lodge Number 1 and the Companions of Columbia Chapter Number 1, Victoria, British Columbia; and to the Brethren of Admiral Lodge Number 170, Salt Spring Island, British Columbia.

*And I will give thee the treasures of darkness,
and hidden riches of secret places.*
—Isa. 45:3 (King James Version)

Contents

Foreword	ix
Acknowledgments	xi
Introduction	xiii
Author's Note	xxv
A Dictionary of Western Alchemy	1
Notes	97
Bibliography	103

Foreword

A renewed interest in the esotericisms of an earlier time seems to be a trend today. In academic circles we've seen everything from Frances Yates's fascinating but lamentably flawed hermetic revival to Richard Kieckheffer's forming of the scholarly *Societas Magica* to Christopher Lehrich's methodologically biased condemnation of Renaissance magical thought as a mere bricolage.

Within esoteric circles, interest in the past has never been new. Certainly the Golden Dawn and the Aurum Solis are rooted in history. They are hardly alone, whether we're talking about the druidic revival of the 1700s or the renewed interest in Hermes Trismegistus in the 1400s. Given the evidence for the practice of alchemy in ancient Egypt, the Greco-Roman world, and medieval Islam, the flourishing of the art in the Middle Ages already constituted a revival. But regardless of the format of esoteric thought inspiring renewed interest, the drive behind that interest has always been the same: *knowledge*.

Jordan Stratford's thirst for knowledge is vast. His involvement in esoterica ranges from Freemasonry to modern Paganism, the Golden Dawn, Gnosticism, and several heretical stops in between. It is inevitable that alchemy, as one of the major sources of Western esoteric thought, would become a focus of his attention.

Alchemy is all about beginnings. It is about middles and endings, too, but it is especially about beginnings. The perfection of the work, the *summum bonum*, is the search for the first material–the *prima material*–from which creation sprang. It is no surprise that the symbols of Christianity, replete with the alpha and the omega, melded so well with the alchemy of medieval and Renaissance Europe.

The act of learning itself is a beginning, and so dictionaries, with every entry a chance to learn something new, are books filled with beginnings. Thus they are a veritable alchemy of words. They are also books *for* beginners. By this I do not necessarily mean readers new to their own vocation or art, though they are certainly included. Rather, I

Foreword

refer to those with a beginner's heart and a beginner's mind. This is the attitude always required for learning, and students who can embrace it have the chance to see the world around them with new eyes. For alchemy, especially the speculative alchemy Jordan discusses, this approach is of great importance.

All disciplines have their specialized, technical language. Most of them have corresponding dictionaries, encyclopedias, or handbooks. Not so the ancient alchemists. To be sure, their language might have been easy to understand if you were a thirteenth-century scholar conversant with the terminologies of metallurgy, medicine, astrology/astronomy, and chymystry. Knowledge of Latin and Greek and an intimate familiarity with both the Bible and Greco-Roman mythology helped, too. After a lengthy apprenticeship, you would be good to go. Today, however, while there are still laboratory alchemists and some of them do take apprentices, you are largely on your own. Thus the need for a dictionary. Luckily, the ancient alchemists left us plenty of words—in fact, many more so than they did diagrams, tables, or convenient photographs of what they were doing.

There are other dictionaries of alchemy, so why read this one? For one thing—steeped in Renaissance lore, Gnostic thought, and Jungian theory as he is—Jordan brings his unique insight to bear, and the reader gains the benefit of this seasoned perspective. For another, the work you see here represents long hours of research and the poring over of dozens of medieval alchemical texts to present the best information possible. But, and perhaps most importantly, the reason to read this dictionary is that it will allow you to see the world with new eyes.

Yes, it will provide you with definitions of alchemical terms. Yes, it will show you different alchemical pictographs and tell you what they mean. It is, after all, a dictionary. But it will also lead you to unexplored ideas. In that way, and as you continue to progress in the Great Work, you will begin to find pieces of your own *prima material*.

–Jeffrey S. Kupperman, Ph.D.
Publisher, *Journal of the Western Mystery Tradition*
February 2011

Acknowledgments

It is entirely possible that you, the reader, may have contributed to this book in some way. Perhaps you made a comment on an online forum that introduced me to a resource, or suggested an etymology, or simply presented me with an unfamiliar term that I later sought to represent here. However you may have done it, I am grateful.

Grateful, too, am I for the countless unnamed librarians, scholars, and webmasters—with a particular nod to Adam McLean—who have posted scans of primary sources to the Internet, making scholarship possible even here on my tiny island in the Pacific. Thank you.

Thanks also to Dr. Jeffrey Kupperman; my fearless editor, Martha Woolverton; Miguel Conner; Xochi Adame; T. Thorn Coyle; the Hermetic Research Society; my bishop, Shaun McCann; Dr. Lisa Lipsett; Dr. Ahava Shira; Robert Birch; and my muse in this as in all things, my brilliant and talented wife, Zandra Stratford.

Introduction

Wisdom's Lacy Hem

*I had discovered, early in my researches, that their
doctrine was no mere chemical fantasy,
but a philosophy they applied to the world, to the
elements, and to man himself.*
—W. B. Yeats, *Rosa Alchemica*

The lunar eclipse. Warmed by spiced rum, we chase glimpses through holes in winter clouds at our diminishing Selene, a disc etched away by some dark vitriol. A bow, a sliver, still of the purest white. And then, still very present, but transmuted into blackest night: the ultimate point of its *nigredo*.

It is the winter solstice. My wife and sons and I crane our necks from the stone porch, in the middle of five acres of evergreen and arbutus forest, on a tiny, windswept Pacific island. It is the first total lunar eclipse occurring on a winter solstice since 1638. And I'm so aware, at this moment, of those earlier moon gazers and how they chose to see the world, to ask deeper questions of their universe than had been asked since the ancients—since Plato and Aristotle.

Then, seemingly without quantifiable transition and yet breathlessly, slowly, an amber haze in the sky appears where the moon should be. As the wind comes up and sweeps the winter stars sharp and clean, the moon is revealed, full and perfect, first carnelian, and then the color of crushed cinnabar or perhaps dragon's blood. *Rubedo*, the completion of the Great Work.

The moon's blooding is every sunrise and sunset on earth, projected onto the lunar surface from the sun's rays. It is a magic lantern show illustrating the clockwork of our universe—the simplest of science and the most miraculous of all things, at once observable. It is a truly alchemical moment.

Introduction

Some have declared that it lies within our choice to gaze continually upon a world of equal or even greater wonder and beauty. It is said by these that the experiments of the alchemists are, in fact, related not to the transmutation of metals, but to the transmutation of the entire universe. This method, or art, or science, or whatever we choose to call it, is simply concerned to restore the delights of the primal paradise; to enable men, if they will, to inhabit a world of joy and splendor. It is perhaps possible that there is such an experiment, and that there are some who have made it.
—Richard Rolle de Hampole, 1380[1]

I recognize the stone as soon as I see it, across the vaulted catacomb. Eighteen months before the eclipse, I'm in the Musée Cluny in Paris. The Cluny is the dream attic of every medievalist—the resting place of every odd "bit" from every cathedral and chateau in the city that no longer has room for it. And among these curiosities is this immediately compelling artifact: the tombstone of fourteenth-century alchemist Nicolas Flamel. Flamel and his wife, Perenelle, had come from modest beginnings to count themselves among the most generous philanthropists of medieval Paris. They attributed this vast wealth, from which they founded and supported numerous churches and hospitals, to an amount of gold produced via the philosopher's stone—the veritable holy grail of the alchemists.

At first, the tombstone's artful symbolism and antiquity of style shout alchemy at the observer. Given the centuries of mythic accretion, this is hardly surprising. Surely we can expect to see the cryptograms of alchemical cipher here—grotesque dragons, hermaphrodites, swans and toads, the sun and moon erupting from exotic plants, a sage concocting homunculi in a jar . . .

But no. What we find instead are Christ, flanked by Peter and Paul, the mundane depiction of sun and moon, and the shrouded figure awaiting judgment and resurrection—these images that are all typical of a pious, accomplished Christian of the fourteenth century. In this

Introduction

engraving of his own device, Flamel chose to be remembered in humility: in his faith, and not in his occult triumph.

Alchemy is not what I wanted it to be. And yet, it is still very much something.

Alchemy (Eg. *al-kemi*) means "of khem," meaning the "black" land of Ægypt. (Stick a shovel in the dirt, and turn it over. If the earth is black, it's fertile and close to the river; you can plant something here, and it will grow and you will not starve to death. If the earth is red, you're in the desert, and if you plant something here it will not grow, and you're going to die. The black land, khem, is what will sustain and nourish you.) Alchemy is both the natural science of that particular culture and the art of working the fertile soil of myth and symbol. It starts from a worldview that connects the ripening of plants with the nature of the names of God—a reaching out to organize a natural history rooted in the Divine.

Egyptian science—astronomy, metallurgy, mathematics, biology, botany, anatomy—existed as a means of proving a connection, a communion, between the finite and the infinite. It's what Catholic writer Andrew Greeley calls "the catholic imagination"—a divinity present, immanent, indwelling, and informing.

Alchemy is a natural philosophy with a practical application. The world of the alchemist makes sense; it is cyclical rather than linear and is largely concerned with *understanding* as a means of healing. The true alchemists were physicians trying to wrap their heads around "the rules" in order to heal a fractured cosmos. It is ultimately a catholic (universal), immanent worldview.

The appeal of the alchemical *weltanschauung* is familiar to any five-year-old with a screwdriver and an alarm clock. What's inside the universe? What is it made of? What makes it work? How do these components—substances, compounds, admixtures, elements—interrelate? And what *idea* makes the whole greater than the sum of its parts?

The *parts* of alchemical study are simple enough: metals (gold, silver, copper, lead, mercury, tin, iron); the visible planets; the elements (earth, air, water, fire, æther); and the processes through which these things can be placed—decay, purification, perfection. As Plato says in *Timaeus*,

Introduction

> Now that which is created is of necessity corporeal and visible and tangible—visible and therefore made of fire—tangible and therefore solid and made of earth. But two terms must be united by a third, which is a mean between them; and had the earth been a surface only, one mean would have sufficed, but two means are required to unite solid bodies. And as the world was composed of solids, between the elements of fire and earth God placed two other elements of air and water, and arranged them in a continuous proportion.
>
> —**Plato**, *Timaeus*[2]

All of this philosophy is in the Hermetic pursuit of "as above, so below"; by observing and understanding the interplay of these substances and forces, we can come to know, and hopefully to master, our own humanity and its circumstances.

It was not just the *chrysopoeia*, the transmutation of base metals into gold, which the alchemists sought. It was also the Universal Medicine, the Elixir of Life—immortality—which beckoned, one further process or refinement away. So, too, they pursued the Reunion, the mystical reconciliation of spirit and matter, of earth and heaven, of man and God.

The assumption stands that alchemy is something that can be learned, which means it is something that is knowable and comprehensible. One is reminded of J. K. Rowling's Snape offering the promise that "I can tell you how to bottle fame, brew glory, and even put a stopper in death." But we are left with the distinct impression that whatever great promises the study of alchemy once held, these are somehow no longer available to us. It seems to me that I am left to consider four options:

1. Alchemy worked, and it still works. I just don't know precisely how, despite countless hours poring over library scans of late medieval manuscripts armed only with some schoolboy Latin, halting New Testament Greek, and the intermittently decipherable notes that eventually became this book.

Introduction

2. Alchemy worked, and it works no longer. Some operating force is not currently active in the world. Which is to say, that the laws of physics have somehow been altered.

3. Alchemy never worked, but the alchemists didn't know that. Those researchers diligently embarking on the effort were misinformed or mistaken in their assumptions. It was a sincere but naive undertaking.

4. Alchemy never worked, and the alchemists knew it. The Great Work was essentially an advertising gimmick to acquire what amounted to industrial chemistry contracts with wealthy patrons.

Certainly, as in any human endeavor throughout history, fraud was a factor. But how to explain the great and sudden wealth of Flamel, a modest scribe whose generous alchemical yield went on to fund his philanthropic efforts? Or Helvetius, who used a "crumb" of the philosopher's stone to make a small amount of gold from lead, with no apparent motive other than curiosity? While these phenomena may belong to the third possibility, some kind of transformation must have taken place that was sufficient to deceive (assuming that the results were not in fact genuine) metallurgists and assayers at the time. So what was that? We do seem to be left with the conclusion that, regardless of whether base metal was turned to gold or to something that could convincingly impersonate gold, *something* occurred that we presently seem unable to replicate.

If the font of that something has run dry, as it appears to have, then why bother with the enterprise at all? Isn't alchemy merely a now-discredited pseudoscience?

Perhaps, although I disagree rather vigorously. Practical alchemy is merely an antique synonym for chemistry. All the great scientific contributions to chemistry made before the modern era—the discoveries of oxygen, hydrogen, phosphorus; the elemental nature of gold, lead, antimony; and the hardening of metals—were made by individuals such as Roger Bacon and Sir Isaac Newton, who referred to themselves as alchemists. On a practical level, there is no difference between the terms *alchemy* and *chemistry*.

Introduction

On a more philosophical note, however, the alchemists bring a quintessence to the modern chemical laboratory. They lived in what we now see as a haunted world—a world in which God, or a "ghost in the machine," was present and sought to make itself known. This fifth element can be understood as *meaning* itself, and, in spite of a current scientific environment that hangs its hat on nihilist materialism while ironically demanding we take the virtues of such a view on faith alone, *meaning* may yet have a contribution to make to contemporary scientific culture.

Speculative Alchemy

This is not a history book; while I concede that the era of the practical alchemist is for the most part behind us, there is still much, I believe, to be gained from inhabiting the alchemical worldview for those engaged in personal or "inner" work.

My love affair with alchemy, like that of Jung, began with the Gnostics. Gnosticism is a philosophy and a literary genre that began in 200 BCE in Alexandria created by diverse authors of Hellenized Jewish origin. They contributed fantastic works of theology and cosmogony, inverting the familiar tales of Greek and Jewish mythology and deconstructing and reinterpreting patterns like jazz musicians. They were heretics and parodists; the punks and beat poets of the ancient world. Gnostic literature addresses themes of memory, identity, the brokenness of human experience, and liberation through insight (*gnosis*) into the true nature of our core reality: Know thyself.

Upon the advent of Christianity, the Gnostics continued to influence the imaginative landscape of religious outliers; many early Christian texts such as the Gospels of Mary and Thomas and even John are equally Christian and Gnostic, with their esoteric and syncretic leanings. But as an emerging orthodoxy marginalized the influence of Gnosticism, its lessons became increasingly inaccessible.

It was Jung's insight that perceived a thematic continuity between the Gnostics and the alchemists:

Introduction

> As far as I could see, the tradition that might have connected Gnosis with the present seemed to have been severed, and for a long time it proved impossible to find any bridge that led from Gnosticism—or neo-Platonism—to the contemporary world. But when I began to understand alchemy I realized that it represented the historical link with Gnosticism, and that a continuity therefore existed between past and present. Grounded in the natural philosophy of the Middle Ages, alchemy formed a bridge on the one hand into the past, into Gnosticism, and on the other into the future, to the modern psychology of the unconscious.[3]
> —C. G. Jung, *Memories, Dreams, Reflections* 1963

So a certain resonance, found in early Gnostic communities, seems to reverberate through the alchemical tradition and is still available to modern seekers.

Alchemy-as-metaphor is prevalent in our culture; it saturates advertising, political commentary, fragrance, cosmetics, gardening, music, and restaurants. But it is in the personal work that the refinement of lead into gold is most valuable. The alchemical process of separation—dividing and identifying our component elements—is the first step in knowing oneself, in understanding that of which we're made: the earth of our bodies and groundedness; the air of our intellect, language and choices; the fire of our passion and refinement; the water of our emotion, intuition, and receptivity.

This metaphorical approach casts alchemy as a medium for communication. Think of a cartoon telephone: two cans and a piece of string. For the communication to take place, you need to put the message in one can (a node) and have the vibration travel along the string (the medium) and finally emerge from the other can (the final node). Each part of this process is described in alchemy thusly:

Black Art: Salt
Black art is about living in your body—figuring out how it works and what it wants, how to recognize when it's in charge. But it's also about burning away negativity, doubt, fear, and attachment. The body is very good at responding to threats, and it tends to be on a hair trigger. But what gets sacrificed when that trigger goes off?

Introduction

Your body—and the works of the physical self, like your job and your home and your cooking skills and sex drive—is a flower, and as such is subject to soil and season, chance and change and weather. Your salt self is beautiful, but once the petals of the flower fall, you're not gluing them back on again. *Memento mori.*

White Art: Mercury
This brings us to the necessity and role of gnosis, or white art, in alchemical work. You need to figure out who the "you" is in the "know yourself" equation. And you can't just rationalize it or have it explained to you. You have to meet *you* and look yourself in the eye until you can say: *Oh, okay, I get it. It's like that. Now I understand.*

This idea involves viewing *gnosis* less as an object or a goal and more as a kind of suspension medium, like the water in a fish tank.

Red Art: Sulfur
Charis. The state of Grace. Reunion with the *pleroma* (Gk: for "fullness"). The totality. Now think back to that tin-can telephone. The first can is salt; the string is mercury; the last can is sulfur. Now imagine that each can connects via more string to more cans. If the signal is strong enough, it can emerge from multiple tin cans on the other end. Myriad sources, myriad voices, one message. One journey with infinite destinations that are again ultimately one.

So, in effect, the work of the alchemist is to internalize the truth of incarnation: that the Infinite is real and abiding and present. It is this incarnationist theme that is at the root of the Western tradition. The Great Work puts you through the four phases: decomposition (of what are you composed? Let's dismantle you and analyze this); modification; separation (distillation); and union (solution, multiplication, projection). When has this process happened to you? In fact, each phase is currently happening to every part of you. In what ways are you aware of this? Where are you putting the emphasis right now? It is this kind of scientific ("knowing") analytic mindfulness that makes a successful alchemist.

Much of my fascination with alchemy stems from my love of myth, my sense of continuity in what we call the Western mystery tradition, and the conviction that divinity abides in the experiential world. That is,

Introduction

1. Alchemy underscores the significance of myth, symbol, and metaphor over concrete histories, events, and personages. It is about putting the center of gravity on subjective experience rather than on objective epistemology. It's important to remember that we are personally and culturally carved out of a solid block of story: Wednesday is the day of Wodin, March is the month of Mars. Christmas trees; green beer on St. Patrick's Day; trick-or-treating; Star light, star bright—all such customs mirror our interplay with myth that is ubiquitous and ambient and, I suggest, necessary.

2. Alchemy shines a light on the continuity of narrative from ancient Egypt through the Greco-Roman world, Hermeticism and Middle Platonism, early Christianity, Gnosticism, medieval Christian mysticism and Jewish Qabalah, and the Renaissance Hermetics and Neoplatonists, through to Jung and the modern era. None of these movements fell out of the sky. They are all part of a contiguous stream of the Western tradition. Alchemy has touched each of the players in this pageant in some way.

3. Alchemy invokes Greeley's "catholic imagination." God is right here, right now. And this God is not William Blake's "old Nobodaddy,"—from his poem of that name—who is a jealous and vengeful character, but a "God" who is that verb, that relationship, that which describes how we connect with things greater than our limited biology. And that connection is as close to us as blades of grass, as motes of dust. There is an inherent and inalienable magic in all things—the transformative magic of idea and ideal.

The creation of this book was serendipitous—which is to say that, at the outset, I had no intention of composing a book. I merely began by taking notes of what I found in alchemy's lovely, quirky, uncooperative manuscripts. I would jot down a symbol to uncover its significance later. I would record a term and later fill in its meaning and etymology. The latter I find particularly instructive in my own reflection and internalization of the entire alchemical gestalt. For example, the word for the process *precipitation*, which describes the entire alchemical process, for example, carries the meaning "head first, falling headlong, head-over-

Introduction

heels." I reflect on how I've had this experience; about how we fall in love, first from our ideal, then visually, then with a lump in the throat and butterflies in the stomach, and then with the glowing of genital coals. There are these little gifts buried in our language, and we invoke them—consciously or unconsciously—when we encounter them.

So, rather than offering a historical overview of alchemy as a phenomenon, or a practical how-to manual, these pages comprise what amounts to the personal lab notes of a sincere and intensely curious seeker. I hope that they will aid your own discovery of and relationship with the primary source texts, which are now widely available.

Magic rites, primitive civilizations, alchemy, the language of flowers, fire, or sleepless nights, are so many stages on the way to unity and the philosophers' stone.
—Albert Camus[4]

The day after the eclipse, we harvest sea salt off the long red dock, the islands parading past in the icy wind. We glug up the gallons of sea water in glass jugs of antique design; this is "the matter" in alchemical parlance. My hands are numb and white from being submerged in winter's tide.

Back home in the kitchen, the vessel we employ to distill the salt is a one-gallon stainless-steel pasta pot. Filtered through cheese cloth, the water rises in its own tide.

The fire comes from a simple gas stove—the sound of the gas catching fire, the electrical snap, the thump of air pressure, and the whoosh of the ignition are familiar hearth sounds.

These three elements—matter, vessel, and fire—comprise the simplest of alchemical workings. The water is brought to a rolling boil to kill off biological contaminants, and the pot is left to steam away our slice of the Pacific. After a few hours, crystals begin to form in the remaining water, and, separating out, they form flakes of the purest white. The salt is wild and pure and honest, and, in its own way, miraculous.

Introduction

Magic rites, primitive civilizations, alchemy, the language of flowers, fire, or sleepless nights, are so many stages on the way to unity and the philosophers' stone.
—Albert Camus

The title of this introduction is "Wisdom's Lacy Hem," an allusion to the alchemists as children, clinging to their mother's skirts. But "lacy hem" is an anagram for "alchemy," and this kind of wordplay is ubiquitous in alchemical literature. Punning and allegory, cipher and anagram, metaphor and poetic vision: These are as significant in the mythic language of alchemy as any chemical formula. I include these here because it seems to me impossible to truly understand the primary source texts through definition alone. A dictionary must address definitions, but the alchemist's perspective must include the Dionysian alongside the Apollonian. Dreaming and intuition accompany observation and testing here. If the introduction as whole seems somehow out of place in a reference book, it is for this reason.

Author's Note

This is not a book *about* alchemy; it is a work *of* alchemy.

Let us imagine that alchemy is a landscape; a territory. Alchemical manuscripts can be seen as maps of this territory, illuminated by travelers from antiquity to the early modern age. In this regard, this book is a compass, a tool for orienting the new traveler with the map in order to better explore the terrain. It serves to be a resource for those actually undertaking the Work itself.

As in any serious field of study, there are those who focus on primary sources and those whose interest is occupied by criticism—the study of the study, if you will. This book was designed to be of benefit to the former; it will hold little interest for the latter. Those who have familiarized themselves with later or secondary works and authors will note that their mention here is omitted. Certainly a history of alchemy or a cultural criticism of alchemy as a phenomenon should include them. This is not such a book. Rather, I have included only those authors and texts the reader is likely to encounter while exploring primary source material.

The source for biographical data has been predominantly *Britannica*, except where there has been contradiction with *The Catholic Encyclopedia*. In these instances I have deferred to the latter, on the assumption that it would have better access to Church records. My guide and constant companion has been Mackey's *Encyclopedia of Freemasonry* of 1917. As to the brevity of biographical entries, I remind the reader that this is a dictionary and not meant to serve as an encyclopedia.

As for curiosities of spelling, much of the alchemical corpus was composed before the standardization of English, and even today there are variations among transliterations from non-Latin alphabets. Even Latin terms will acquire certain deviations from the norm in alchemical texts. I have chosen the most commonly encountered presentations, not always the most correct. Likewise, with the symbols included, I have gone to the earliest sources, then tempered my selection in favor

Author's Note

of those most often employed. The symbols used here are in no way meant to be considered comprehensive; they are just those the modern researcher is most likely to face (or something similar). It must be kept in mind that the main function of these symbols was to occlude, and individual alchemists would substitute a common sigil with one of their own device or deliberately misuse another cipher in order to mislead a rival. The whole undertaking is something of a puzzle and requires us to have our wits—both in the sense of intelligence and humor—about us.

A Dictionary of Western Alchemy

The alchemists in their search for gold discovered many other things of greater value.
—Arthur Schopenhauer

ablation
The separation of components in a vessel by skimming to remove the top layer. From Latin *ablationem*, "taking away."

ablution
Purification by washing with water. From Latin *abluere*, "to wash off."

acetabulus
A small cup used for vinegar. From Latin *acetum*, "vinegar."

acetum
Literally, Latin, "vinegar." From Latin *acidus*, "sour."

acetum philosophorum
A vitriol, likely *aqua regia*. From Latin, literally, "vinegar of the philosophers."

acid
A compound that transfers a hydrogen ion to another compound; a medium for breaking down a substance. From Latin *acidus*, "sour."

adamas
Unbreakable. Literally, Greek, "diamond."

aer
Greek, "air."

aerugo
Verdigris, from copper. From Latin *aes*, "copper."

aes
Brass; also used generically for any common metallic substance, such as copper or bronze. Literally, Latin, "copper."

aes cyprium
Copper. Literally, Latin, "brass of Cyprus."

aether
The hypothetical element that acts as a suspension medium of the celestial realm. In Greek religion, aether is the air breathed by the gods. Plato classifies it as the quintessence, or fifth element. From Greek *aithein*, "to shine."

agate
A semiprecious stone of quartz used for its resistance to acids and often employed in certain instruments, such as a mortar and pestle. Named for the Greek river Achates.

Agathodaemon
A (possibly apocryphal) third-century alchemist of Roman Egypt, recalled in later texts for his work with arsenic and silver.

air
The first and supreme of the four elements of classical philosophy, air was regarded as the *arche*, or first principle (Greek), of creation. Plato's *Timaeus* identifies elemental air with the Platonic solid of the octahedron, and Aristotle defines air as having the qualities of hot and wet—distinct from aether, which was immutable and ideal. However, the

relationship between alchemical "air" and our atmosphere is not a direct and literal one, as A. E. Waite notes:

> Eugenius Philalethes says that the air is not an element, but a certain miraculous hermaphrodite, the cement of two worlds, and a medley of extremes. It is the sea of things invisible, and retains the species of all things whatsoever. It is also the envelope of the life of our sensitive spirit. The First Matter of the philosophers is compared to air because of this restlessness.[1]

Throughout Western cultures, air is associated with spirit and the divine, both in Latin (*spiritus*, "breath") and in Greek (*pneuma*, "spirit"), the latter invoking the Egyptian *khnum*, ("breath"). In magical practice air corresponds to the East, to the suit of swords in the tarot, to intelligence and language, and to the Archangel Raphael. From Greek *aer*.

 alabrot
Potassium nitrate; saltpeter. Possibly from Arabic, literally "sweet salt."

 **albedo**
Whitening, the second stage of the Great Work, arrived at through the burning away of impurity and symbolized by a swan. This phase is associated with lunar forces and influence. In optics this term refers to the ratio of incident light reflected by a surface. From Latin *alba*, "white."

 albumen
Egg white. From Latin *alba*, "white."

Albertus Magnus (c. 1200–1280)

A scholar, philosopher, saint, and doctor of the Church, Albertus applied Aristotelian thought to Roman Catholic teaching. Born in Bavaria and educated in Padua, he entered the Dominican Order in 1221 or 1223 and received his doctorate in Paris in 1245. In 1260 he became the bishop of Regensburg. He was the mentor of Thomas Aquinas, who reported that Albertus witnessed the transmutation of base metal into gold. His extensive alchemical writings are collected under the title *Theatrum Chemicum*, but later texts such as the *Secreta Alberti* and *Experimenta Alberti* are likely pseudepigraphic. Even during his lifetime, Albertus was hailed as *Magnus* ("The Great") by his contemporaries, including the alchemist Roger Bacon.

albification
See *albedo*. From Latin *alba*, "white."

alchemy
The philosophical and natural science of elements and interactions. Since the Enlightenment, alchemy has been associated with charlatanism and greed, based on a literal interpretation of the desire to turn lead into pure gold. However, the main reward sought by the authors of alchemical texts seems to have been knowledge, both of the natural world and of the place of the Divine within it. In the words of Paracelsus,

> When the Philosophers speak of gold and silver, from which they extract their matter, are we to suppose that they refer to the vulgar gold and silver? By no means; vulgar silver and gold are dead, while those of the Philosophers are full of life.[2]

Grounded in a worldview informed by both Hermetic and Neoplatonic influences, the primary activity of alchemists involves three pursuits, each with a practical and a spiritual aspect:

1. the Great Work, being the transmutation of base metals into gold by means of the *lapis philosophorum*, the philosopher's stone;

2. the concoction of the elixir of life, a universal medicine; and

3. the reintegration, a reconciliation between spirit and matter.

However, as Jakob Boehme states in *De Signatura Rerum*, "There is no real difference between Eternal Birth, Reintegration, and the discovery of the Philosopher's Stone. Everything having issued forth from Unity, all must return to it in the same manner."[3]

Until the early modern era, alchemy was indistinguishable from chemistry, and alchemists were among those who developed the scientific method.

From Arabic *al-kemi*, "of Egypt," and *khem*, referring to black, fertile soil.

alcohol
An organic compound in the form of a clear, flammable liquid, most commonly ethanol, produced through fermentation or distillation. Its use in English to refer to any distillate dates from the seventeenth century. From Arabic *kohl*, "stain" or "paint," referring originally to antimony.

alembic
A simple distillation apparatus comprised of a retort (a spherical vessel with a long, downward-sloping tube) and a cucurbit (the receiving vessel). A substance is heated in the retort, condensing at the capital and collecting downward into the cucurbit. It is used as a metaphor for rigorous intellectual and spiritual inquiry, distilling only the essence of an idea. From Greek *ambix*, "mixing cup."

alkahest
A theoretical universal solvent, in which any substance would be reduced to its elemental form. Paracelsus developed such a mixture of caustic lime, carbonate of potash, and alcohol. A pseudo-Arabic word coined by Paracelcus.

alkali
A caustic salt of an alkaline earth-metal element, most commonly magnesium or calcium. From Arabic *al-qaliy*, "the calcinated ashes."

alkanet
Dyer's bugloss (*Alkanna tinctoria*), a borage plant used to make red dye. From Arabic *al-kanna*, "henna."

aludel
A pear-shaped earthenware tube, also called "the Hermetic vase," used for condensation at the end of the sublimation process. Because of its association with completion of the operation, it is often used as a metaphor for personal realization or achievement. Referring to the shape of a lute, from Arabic *ud*, "wood."

alum
Used to describe a range of compounds, most commonly iron sulfate, used in dyeing. From Greek *aludoimos*, "bitter."

amalgam, amalgamation
The product and process of forming a "soft mass" via the reaction of mercury with another substance, most commonly silver or gold. From Greek *malakos*, "soft."

amber
Fossilized tree resin, believed to prevent infection. In Greek, *amber* is "elektron," noted for its electrostatic properties. From Arabic *anbar*, "amber."

amethyst
A purple quartz used since antiquity in the belief it prevents drunkenness. From Greek *a-*, "not," and *methustos*, "intoxicated."

ammonia
A compound of hydrogen and nitrogen, also referred to in alchemical texts as "volatile alkali." Ammonia is named for the deposits of ammonium chloride located in Libya next to the Temple of the god Amun, the (eventual) supreme creator deity of the Egyptian pantheon.

amphora
A two-handled ceramic vase with an elongated neck. From Greek *amphi-*, "both sides," and *pherein*, "to carry."

animals

Certain animals, both real and fantastical, are frequently depicted in alchemical texts. Often they are used to signify phases of the Great Work: the toad and the crow (black phase); the eagle, the unicorn, and swans (white phase); and the pelican and the cockerel (red phase). Other commonly depicted animals include the peacock, the basilisk, and the lion, in various colors.

anise

Pimpinella anisum; a flavorful, flowering plant used to combat flatulence.

anneal

In metallurgy, to treat with heat to change the properties of a material or to make it easier to manipulate. From Old English *aelan*, "to burn."

antimony

Stibium; a brittle, silver-white crystalline element. Possibly known to Geber in the eighth century, it was first identified in Europe in the sixteenth century by the Italian metallurgist Vannoccio Biringuccio. Antimony was symbolized by a grey wolf because of the propensity for molten antimony to "overcome" other elements by quickly consuming them into alloys. Latinization of the Greek *stimmi*, from ancient Egyptian *stm*, "antimony."

aqua fortis

A solvent made from saltpeter (potassium nitrate), used for dissolving silver. Literally, Latin, "strong water."

aqua regia

Nitro-hydrochloric acid, a yellow, highly corrosive solvent capable of dissolving gold and platinum. Geber discovered it in the eighth century by combining salt with sulfuric acid. Literally, Latin, "royal water."

aqua pluvialis
Latin, "rain water."

aqua vitae
A general term for all distillates, originally applicable to a concentrated aqueous solution of ethanol. A direct translation from the Latin to Gaelic yeilds *uisce beatha*, the origin of "whisky." Literally, Latin, "water of life."

archaeus
The inner or hidden nature of a substance. From Greek *arkos*, "secret" or "that which is locked away or protected."

archemy, archimastry
A neologism that implies superior practice of the art of alchemy. It seems to be used by some alchemists to differentiate their activities from those of their rivals (Giovanni Agostino Pantheo refers to the rival practice as "illicit and duplicitous alchemy" as opposed to his own work published in 1530 as *Voarchadumia*, "the cabala of metals"). From *Norton's Ordinal* (1477), comes the quote, "Mastery full marvelous and Archimastry is the tincture of holy alchemy." By the nineteenth century, however, *archimie* had become a French term for chemical analysis in metallurgy.

arena
Common sand, used in filtration. From Latin *harena*, "sand."

 argentum
Latin, "silver."

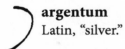 **argentum vivum**
Elemental mercury; quicksilver. Literally, Latin, "living silver."

 arsenic
A crystalline chemical element known for its poisonous properties. It can be categorized as *album*, *citrinum*, or *rubrum* for white, yellow,

or red, respectively. From Persian *zarnikh*, "yellow orpiment," a yellow mineral found in hot springs.

art royal
A euphemism for alchemy.

asbestos
A set of silicate minerals known for their heat- and fire-resistant properties. In some texts the word is used to describe lime. Legend attributes it to the wool of salamanders (fire spirits). From Greek *asbestos*, "unquenchable."

ascension
The rising of the active component to the top of the vessel. From Latin *ascendere*, "to climb."

Ashmole, Elias (1617–92)
A famed English intellectual, astrologer, and alchemist and the founder of the Royal Society. His collection of natural history artifacts forms the origin of the Ashmolean Museum in Oxford. An artillery captain (due to his gift of mathematics), Ashmole served on the Royalist side of the English Civil War. After the Restoration he served the Crown as auditor and herald and held numerous other posts. Ashmole makes one of the earliest references to freemasonry and was possibly a Rosicrucian. He became a doctor of medicine in 1669 and published a collection of alchemical works, the *Theatrum Chemicum Britanicum*.

assation
The process by which a substance is reduced to ash, effectively by roasting. From Greek *azein*, "to dry."

athanor
A coal-burning furnace, also called "the tower furnace," noted in alchemical operation for its slow burning and infrequent stoking. Philostratus mentions an "occult hill" called Athanor in his *Life of Apollonius* (c. 200 CE), referring to Apollonius of Tyana, the neo-Pythagorean philosopher.

atramentum
Vitriol; any black liquid such as ink or dye. From Latin *atrox*, "cruel."

aurichalcum
A golden-toned alloy, possibly bronze, or of gold and copper. From Greek *oros*, "mountain," and *chalkos*, "copper" or "bronze."

aurum
Latin, "gold."

azoth
An ideal formula; the completion or union of opposites in mastery. Generally associated with the element mercury, azoth is frequently depicted by the caduceus. Alternatively "azote." From Arabic *al-zauq*, "mercury."

B

Then circulate them so thou shall
To heale in man diseases all
For then thou has Electrum right
The first essence of the Sonne bright
This is the Philosophers Sulfur vive
Theire Tinctur, lead, theire Gold of life
—Edmund Dickinson, 17th Century

Bacon, Roger (c. 1214–94)
One of the pioneers of experimental science, an English Franciscan with an encyclopedic knowledge of physics, astronomy, optics, philosophy, theology, and alchemy. A master of Aristotle at Oxford, Bacon is thought to be the author of the *Speculum Alchemiae* and possibly of the famed Voynich manuscript.

bain-marie
A double-hulled pot used for warming components slowly and evenly. The outer pot is subjected to direct heat and is filled with a liquid, usually water, into which the second, inner pot is placed. The substance to be heated is placed inside the second vessel. The bain-marie (literally, French, "Mary's bath") is named for "Mary the Jewess," the legendary alchemist who is sometimes associated with the sister of Moses or Mary Magdalene.

balsam
An emulsion serving as a suspension medium for essences. Paracelsus considered the balsam of vitriol ("the green lion") to be the vital

essence of the human body, repelling putrefaction. From Arabic *basham*, "perfume."

baryte
Barium sulfate, noted in alchemy for its phosphorescent properties. From Greek *barus*, "heavy."

basilisk
A fantastical, venomous beast with the head of a cock and the body of a lion or sometimes a dragon. The basilisk is the *prima materia*, the chaotic base matter from which all organized matter subsequently derives. From Greek *basiliskos*, "little king."

Becher, Johann Joachim (1635–82)
A physician, metallurgist, encyclopedist, and alchemist, Becher was a German scholar who first proposed the idea of phlogiston. He proposed an invented, universal language and contributed 100,000 words to this effort. He also aspired to concoct an invisibility potion.

beeswing
Potassium bitartrate, the crude form of cream of tartar.

bezoar
An aggregate mass formed by the clumping of certain compounds. Such masses found in nature, like balls of hair, root tangles, et cetera, were considered to have magical healing properties. From Persian *padzahr*, "counter poison."

bile
A dark yellow fluid produced by the liver, thought to regulate depression and aggression. From Latin *bilis*.

birds
Used to represent phases of alchemical process and spiritual experience: the black crow (*nigredo*), the white swan (*albedo*), the peacock (transitional iridescence), the pelican (sacrifice), and the phoenix (completion, resurrection).

bismuth
A heavy white element similar to arsenic and antimony. In 1980, a microscopic portion of bismuth was successfully turned into gold by the use of nuclear physics. Possibly from Old High German *hwiz*, "white."

bitumen
A sticky, black, tar-like form of petroleum used in sealants and waterproofing. From Celtic *betu*, "birch resin."

black art
A euphemism for alchemy, derived from its purported Egyptian origins (Arabic *khem*, "black," as in the black, fertile soil of Egypt). Sometimes used to describe the *nigredo* (blackening phase) of the Great Work.

black lion
Caput mortuum.

Boehme, Jakob (1575–1624)
A German theologian, visionary, and philosopher. Boehme's writings, influenced by his study of Paracelsus, contain alchemical and qabalistic thought.

boiler
A closed vessel for heating liquid.

borax
A crystalline mineral salt that dissolves easily in water, borax is used as a flux in the soldering of gold or silver. From Arabic *buruq*.

botarion
A cucurbit, a vessel used in digestion. Possibly from Greek *bathos*, "cavity" or "depth."

Boyle, Robert (1627–91)
Widely regarded as the father of modern chemistry. Boyle's work was firmly grounded in the alchemical world. A wealthy member of the

famed Invisible College (which later became the Royal Society of London), Boyle subscribed to the practical possibility of the transmutation of metals and worked successfully to overturn the statute against the alchemical multiplication of gold. However, he rejected the historic framework of alchemical understanding, such as the prime importance of sulfur, salt, and mercury.

Brand, Hennig (c. 1630–c. 1710)
A German alchemist credited with the discovery of phosphorous, which he extracted from urine.

brass
Originally, an alloy of copper and tin (which we know as bronze); later, an alloy of zinc and copper; any crude metal dug from the earth; ore; anything unrefined.

brick dust
A substance used in filtration, represented by the initials "fl" from Latin *farina laterum*.

brimstone
Sulfur. From Old English *brin*, "to burn."

bronze
An alloy of tin and copper. Possibly from Persian *birinj*, "copper."

butter of antimony
Antimony trichloride.

butter of tin
Stannic chloride.

The Masters of nature have written much that the art was not to bee perfected constantly, on purpose, that the unwise might reach to it, but to the just and to the godly it becomes profitable both here and hereafter.
—Arnaldus de Villa Nova,
A Chymicall Treatise

Cagliostro, Alessandro (1743–95)
The nom de guerre of Italian alchemist, philanthropist, and forger Giuseppe Balsamo, Cagliostro was born in Sicily. Assuming the persona of a count, he traveled to Malta, where he established himself as a pharmacist, which gave him means to practice alchemy. He traveled to Rome and Paris, where he was embroiled in the notorious "Affair of the Necklace" with Marie Antoinette and imprisoned at the Bastille. Upon his release, he fled to England, later returning to Rome, where he was imprisoned for being a Freemason.

calamine
Zinc oxide or zinc carbonate; or an alloy of lead, tin, and zinc. From Greek *kadmeia*, "of Kadmos" (an alternative name for Thebes).

calcination
The process of burning or roasting in an open vessel. From Latin *calx*, "limestone."

calomel
Mercurous chloride.

calx
Lime, or quicklime; more generally, any powder derived from calcination. Literally, Latin, "chalk."

camphor
A white, waxy solid derived from the camphor laurel. From Arabic *kafur*.

caput mortuum
The residue of sublimation; *nigredo*; anything leftover or worthless. Literally, Latin, "death's head," a human skull.

cement
The dry powder of any hardening medium used to bind materials together. Early cement was a combination of burnt lime and crushed rock. From Latin *caementa*, "chipped stone."

ceration
The process of softening a material into a waxy consistency. From Latin *cera*, "wax."

chalcite
Copper pyrite (no relation to calcite). From Latin *calx*, "limestone."

charcoal
A black carbon ash created by the slow heating of organic elements in a low-oxygen environment and used in filtration.

chemical wedding
The metaphorical initiation of the alchemist by purification and assaying, representing the reconciliation of opposites as an essential component of the Great Work. Usually associated with the *citrinas* stage of the Work, it is poetically invoked in the 1459 treatise *The Chemical Wedding of Christian Rosenkreutz*, anonymously attributed to Johann Valentin Andreae.

chrysocolla
Copper silicate, a gem similar to turquoise. From Greek *chrysos*, "gold," and *kolla*, "glue," a reference to the mineral's use in soldering gold.

chrysopoeia
One of the properties of the philosopher's stone. Literally, Greek, "to make into gold."

chrystallus
A legendary stone material, the discovery of which was attributed to Aristotle, and which was equated with sapphire, comprising "the great aqueous heaven, the firm abode of all-powerful God." From *crystal*, by way of Greek *krystallos*, "frost."

cibation
The process of adding to the content of a crucible. From Latin *cibus*, "food," implying the act of "feeding" a substance to the vessel.

cineration
Reduction to ash. From Latin *cineris*, "ashes."

cinnabar
The red form of mercuric sulfide, the common ore of mercury. From Greek *kinnabari*, of unknown meaning.

circulation
Purification through distillation and condensation in a closed vessel.

citrinas
Yellowing; the completion of fermentation. From Latin *citrus*, by way of Greek *kedros*, "cedar."

coadunation
The process of uniting by growth. From Latin *ad-* "to," and *unitas*, "unity" or *unus*, "one."

coagulation
Curdling; the process of thickening a liquid. From Latin *co-*, "together," and *agere*, "to drive."

cobalt
A lustrous silver element used to impart a blue color in glass and ceramics. Named for the kobold, the mythical German earth spirit or goblin associated with mines.

coction
Cooking; the process of heating over an extended period; digestion. From Latin *coquere*, "to cook."

cohobation
A kind of circulation in which the distilled liquid of a substance is removed and reintroduced. From Arabic *ke'aba*, "to repeat."

colcotharum
Crocus; iron oxide. Possibly from Greek *Kolkhis*, a region on the Black Sea, and the mythological location of the Golden Fleece.

colliquation
Degeneration into liquid as part of putrefaction; the fusing together of two or more substances by melting. From Latin *liquere*, "to flow."

coloration
The process of manipulating the color of a substance by tinting or dyeing.

combustion
The process of burning. From Latin *comburere*, "to burn."

comminution
Reduction to powder by grinding, most commonly bone. From Latin *minuere*, "to lessen."

composition
The process by which two or more substances are joined. From Latin *com-*, "together," and *ponere*, "to place."

conception
The union of the female and male qualities of a substance. From Latin *capere*, "to take" or "to grasp."

concoction
From Latin *coquere*, "to cook." See coction.

congelation
From Latin *gelare*, "to freeze." See coagulation.

conglutination
The deterioration of a substance into a gluey consistency as a result of putrefaction. From Latin *gluten*, "glue."

conjunction
The process of joining, usually of substances with opposing qualities (base and fine, male and female, etc.). From Latin *jugare*, "to join."

contrition
The reduction of a substance through heating or evaporation. From Latin *terere*, "to rub" or "to grind," as to grind to pieces.

copper
A malleable reddish element used as a conductor. Alchemically, copper relates to the *rubedo*, the spiritual act of union with the Divine. Its name derives from Cyprus, where it was mined.

copperas
A term for copper (blue copperas), iron (green copperas), and zinc (white copperas) sulfates. Derives from Cyprus, where copper was mined.

copulation
Conjunction. From Latin *copular*, "to tie."

corrosion
The process of disintegrating a material substance into its components; the return of an alchemist's work into base states. From Latin *rodere*, "to gnaw," as would a rodent.

cribation
The process of sifting a powdery substance through a mesh or sieve. From Germanic *krebe*, "basket" or "weave."

crocus
Iron sulfate (*crocus martis*) or copper oxide (*crocus veneris*); sometimes *aes ustum* (Latin, "burnt brass").

crow
A symbol for the *nigredo* (blackening phase) of the Great Work.

crucible
A vessel designed for sustaining high temperatures, used in smelting and refining metals.

crystallization
The formation of crystals in a substance in a solution.

cucurbit
The receiving vessel of an alembic. The Latin name for the family of squash and gourds, referring to the shape of the apparatus.

cupel
A pot made from bone ash and used in the refinement of noble metals from base. A variation of *cup*.

cupellation
The process of assaying and refining precious metals. From *cupel*, a variation of *cup*.

MONTH OF Y COLRICK

HERE IS YE LAST OF Y RED AND Y BEGINING TO PVT AWAY Y DED Y ELEXER VITE

THE BEDE OF HERMES IS MI NAME ETING MI WINES TO MAKE ME TAME

This Art, of bringing all Imperfect Metals to Perfection, hath been asserted for Truth, by Men of almost every Degree, in most Ages of the World.
—Eugenius Philalethes,
Concerning the Hermetic Art, 1714

dealbation
The transformation of a black substance into a white substance, as in ash. From Latin *alba,* "white."

decoction
Unaided digestion in a vessel. From Latin *coquere,* "to cook."

decrepitation
The process of cracking or shattering a substance by heating. From Latin *crepare,* "to crack."

Dee, John (1527–1608)
A famed alchemist, Hermetic philosopher, and trusted advisor to Queen Elizabeth I, Dee was a brilliant astrologer, mathematician, cryptographer, and navigator. He is likely the architect of the "British Empire," a phrase he is reported to have coined. Dee's exhaustive work in divination resulted in a comprehensive angelic lexicon and cosmography, and his library was considered to be among the greatest in all of Europe.

deliquium
The process of melting or dissolving into a medium. From Latin *liquere*, "to flow."

descension
The process of going downward or sinking; the fall or return of an ascended substance. From Latin *de-*, "down," and *scandere*, "to climb."

dessication
The process of drying to the point at which all moisture is removed; contraction due to baking. From Latin *siccus*, "dry."

detonation
The process of explosive burning. From Latin *tonare*, "thunder."

de Villa Nova, Arnaldus (1235–1311)
A scientist and Arabist, responsible for the translation of numerous medical and alchemical texts from Arabic into Latin and then into Spanish. He was born in Valencia, of Catalan origin, and his career took him from Aragon to Paris, from whence he was summoned by Pope Clement V (of Templar-crushing notoriety), but he died en route to Avignon.

diaphoretics
Those elements that "sweat" when heated: antimony, copper, gold, iron, lead, silver, and tin. From Greek *diaphorein*, "to sweat."

digestion
The process of exposing a substance to moisture and heat by means of softening or disintegration. From Latin *digerere*, "to break down."

disintegration
The process of breaking up or decomposing a substance into its component parts. From Latin *dis-*, "oppose," and *integrare*, "to make whole."

dispoliaration
The process of dissolving a dead substance; liquefaction. From Latin *despoliare*, "to plunder"; related to "spoil."

dissociation
The process in which compounds such as salts separate into smaller particles, usually in a medium. Dissociation is often reversible by evaporation.

dissolution
The process of mixing a substance into a solvent to form an indistinguishable whole. From Latin *solver*, "to loosen."

distillation
The separation of mixtures through the application of heat, evaporation, and condensation. Distillation is used as a metaphor for a purifying ordeal, the spiritual trials of the alchemist. Hypatia of Alexandria is thought to have invented a distillation apparatus in the fourth century. From Latin *stillare*, "drop."

divaporation
The emanation of dry vapor from a substance. From Latin *vaporem*, "steam."

division
The separation of elements in a substance by various means. From Latin *videre*, "to separate."

dove
A symbol for the transition from the *nigredo* (blackening phase) to the *albedo* (whitening phase) of the Great Work.

drachma
A small unit of weight; a dram. A modern dram is 3.697 ml, or a teaspoon. From Greek *drachme*, "handful."

dragon
A mythical beast used symbolically to represent the *nigredo* (blackening phase) of the Great Work of the alchemist. In alchemical illustrations, the dragon is often depicted swallowing its own tail, as does the ouroboros, representing the cycles of the seasons and the inescapability of the physical world. When the dragon is shown winged, it can symbolize the *rubedo* (reddening phase).

dragon's blood
Any number of dried, red tree resins, most commonly of the genus *Dracaena*; sometimes cinnabar (mercury sulfide).

dung
In alchemical texts, when unspecified, usually from a horse.

E

*Our prepared substance is
much better and more honourable than gold, and has power to do that
which gold cannot do, viz, to change the
common matter of all metals into gold.*
—A German Sage, *A Tract of Great Price*, 1423

eagle
A symbol for the *albedo* (whitening phase) of the Great Work. It is considered to be feminine and in sexual alchemy represents the vagina charged with seminal fluid.

earth
In alchemy, a Platonic element, an ideal of "earthiness," containing the material qualities of dryness and cold and the human characteristics of practicality and materialism. In Plato's *Timaeus*, the earth element is associated with the cube. Alchemically, this element is typified by salt.

ebullition
The process of boiling. From Latin *bullere*, "to bubble."

edulceration
The removal of salts by washing. From Latin *dulcis*, "sweet."

egg
Unless specified, a chicken's egg; a flask in the shape of an egg. Eggs also symbolize fertility, the physical world, potentiality, and, rarely, sulfur.

elaboration
The purification of a substance (or indeed, the alchemist) by means of the Great Work. From Latin *elaborare*, "to produce by labor."

electrum
An alloy of silver and gold, used in coins in ancient Lydia in the seventh century BCE. Because of variable gold content, it was difficult to assess any given coin's actual value. From Greek *elektron*, "amber."

elements
The four Platonic elements that, in the alchemists' worldview, are assumed to compose each substance, experience, or idea: earth, air, fire, and water. These are the "ideal states" of these phenomena, not the mundane or earthly states in which we encounter them. Plato explains in his cosmography, *Timaeus*,

> God placed water and air in the mean between fire and earth, and made them to have the same proportion so far as was possible (as fire is to air so is air to water, and as air is to water so is water to earth); and thus he bound and put together a visible and tangible heaven.[4]

These elements can be understood as having expression in personality: earth is practical, physical, and material; air is intellectual, abstract, and idealistic; water is emotional, intuitive, and receptive; and fire is passionate, willful, and active.

Traditionally, alchemy speaks of a quintessence, or fifth element, the moving or animating spirit dwelling within each of the other elements.

Alternatively, alchemical texts may apply the term *elements* to the three principles of salt (contraction), sulfur (expansion), and mercury (integration).

elevation
The separation of the subtle part of a substance from its base by means of rising in a vessel. From Latin *levis*, "light weight."

elixeration
The process of composing an elixir. From Arabic *al-iksir*, possibly related to Greek *xeros*, "dry."

elixir
Originally a medical powder. Geber first uses the term to describe a substance that can significantly alter the properties of other substances—the philosopher's stone. From Arabic *al-iksir*, possibly related to Greek *xeros*, "dry."

elixir of life
An ideal potion bestowing immortality to the one who consumes it, also known as the quintessence of life. It is the universal medicine, the panacea (so named for the Greek goddess of healing), which is one of the aims of the Great Work of alchemy.

evaporation
The process of removing moisture by application of heat or exposure to a dry environment. From Latin *vaporem*, "steam."

exaltation
The process of purifying a substance to a more noble or sublime nature. From Latin *ex-*, "up" or "out," and *altus*, "high."

exhalation
The emanation of vapor or fumes. From Latin *halare*, "to breathe."

expression
The production of moisture from a substance by pressing or crushing.

extraction
The process of refining a substance by soaking in alcohol and removing residue. From Latin *trahere*, "to draw out."

F

My Child, lock this up in thy heart and understanding, this Saturn is the Stone which the Philosophers will not name, whose name is concealed unto this day; for if it's name were known, then many would operate, and the Art would be common.
—Johann Isaac Hollandus,
A Work of Saturn, 1670

faex vini
A substance used in the preparation of nitrum (ammonium nitrate); yeast. Literally, Latin, "dregs of wine."

fermentation
The anaerobic decomposition of organic compounds, resulting in alcohol, carbon dioxide, and water. From Latin *fermentum*, "yeast."

ferrugo
Latin, "rust"; or rusty coloration.

ferrum
Latin, "iron."

filtration
The processing of a substance through a filter, usually cheesecloth. Latin by way of Germanic *feltaz*, "felt."

△ **fire**
The first of the four classical elements, fire is the ideal force and the manifestation of divine enlightenment, will, and illumination. Alchemically, and later chemically, its glyph is the symbol of transformation. It is the subject of the Promethean theft, and the liberating fire of knowledge. It is also the punitive fire of Hades, and the creative (practical) fire of Hephaistos. It is martial action and passion. Plato identified fire with the tetrahedron and the qualities of heat and dryness. Fire is also associated with the alchemical principle of sulfur and the magical spirit of the salamander. In tarot, fire is represented by the suit of wands or staves, and the angel of fire is Michael.

ℨℨ **fixation**
The process of solidifying a substance. From Latin *figere*, "to attach."

Flamel, Nicolas (1330–1418)
An enigmatic figure who has been posthumously reputed to be perhaps the greatest alchemist of all time, having produced the philosopher's stone with the assistance of his wife, Perenelle. The two donated the results of their work to charity, funding no fewer than fourteen hospitals and seven churches. Flamel is rumored to have acquired a manuscript in Spain that revealed to him the secrets of alchemy. His tombstone is on display in the Musée Cluny in Paris.

flowers (of lead, antimony, brass, steel, copper, sulfur, and vitriol)
Rusts or oxides; preparations made by sublimation of select metals.

Fludd, Robert (1574–1637)
An English alchemist, astrologer, mathematician, and physician, Fludd was one of the greatest occultists of the Elizabethan era. A proponent of Rosicrucian ideals and the author of *An Apologetic Tract Defending the Purity of the Society of the Rosy Cross* (1617), he is best known for having discovered the circulation of blood.

flux
A chemical wash used in the preparation of metals for soldering. Flux removes impurities while preventing formation of metal oxides, preventing solder from beading. Literally, Latin, "flow."

foliation
The process of making "leaves" or foil of a substance, such as silver leaf. From Latin *folia*, "leaves."

froth of niter
The subtle part of *nitrum* (potash or soda).

fulmination
The process of transforming a substance into an explosive, such as fulminate of mercury or of silver. From Latin *fulgere*, "to flash."

fumigation
The process of exposing a substance to smoke. From Latin *fumus*, "smoke."

furnace
The hearth of the alchemist. Furnaces of various types were employed for different functions. In *The Art of Distillation* (1651), John French writes,

> The matter of furnaces is various, for they may be made either of brick and clay, or clay alone with whites of eggs, hair and filings or iron (and of these if the clay be fat are made the best and most durable furnaces) or of iron or copper, cast or forged. The forms also of furnaces are various.
>
> The fittest form for distillation is round; for so the heat of the fire being carried up equally diffuses itself every way, which happens not in a furnace of another figure, as four square or triangular, for the corners disperse and separate the force of the fire. Their magnitude must be such as shall be fit for the receiving of the vessel; their thickness so great as necessity shall seem to require; only thus much observe, that if they be of forged iron or copper, they must be coated inside, especially if you intend to use them for a strong fire. They must be made with two bottoms dis-

tinguished, as it were, into two forges, the one below which may receive the ashes, the other above to contain the fire. The bottom of this upper must either be an iron grate or else an iron plate perforated with many holes so that the ashes may the more easily fall down into the bottom, which otherwise would put out the fire. Yet some furnaces have three partitions, as the furnace for reverberation, and the register furnace. In the first and lowest the ashes are received. In the second the fire is put, and in the third of the furnace for reverberation, the matter which is to be reverberated.[5]

fusion

The process of sealing or combining two or more substances through application of intense heat. From Latin *fusus*, a form of the verb *fundere*, "to melt."

G

*Mingle therefore Iron and Clay, and thou shalt
have the foundation of Gold.*
—Aesch-Mezareph,
or Purifying Fire, 16th Century

Geber (721–815)
Jabir, or Abu Musa Jabir ibn Hayyan al-Azdi, the Arabic father of alchemy. Jabir was a brilliant mathematician, astrologer, engineer, Neoplatonist, and philosopher—the Da Vinci of the Arab world. His contribution to alchemy, and to chemistry, is unparalleled. He developed a scientific method for experimentation, as well as basic fixtures of laboratory equipment, such as the alembic. He is also credited with the discovery of citric, acetic, tartaric, and hydrochloric acids and the development of *aqua regia*, capable of dissolving gold. He also isolated, identified, and purified a number of elements: sulfur, bismuth, mercury, arsenic, and antimony. He named the three principles of alchemy (salt, sulfur, and mercury), and it is Jabir who is the philosopher of the philosopher's stone. At the heart of his natural-science pursuits was the quest for spiritual refinement and union with God. The obfuscation of his experimental records by means of symbol and metaphor gives the origins of the word *gibberish*.

glass
Any inorganic product of fusion that has cooled into a noncrystalline solid; also a synonym for "mirror."

glutination
The process by which a substance becomes glue-like. *See* coagulation.

gold
The most malleable of all the elemental metals, gold has been sought and prized throughout human history. It is a metaphor for itself, representing all that is valued and valuable and being both a symbol for and an embodiment of riches material and spiritual. It was the pursuit of the alchemical secret of gold that financed the studies of the classical and medieval alchemists; while the patrons awaited production of a material substance, the scholars held that it was they themselves who were transmuted into something more precious, more noble. Gold is associated with the qabalistic sphere of *tiphereth* (beauty) and the sun. It is frequently referred to in alchemical texts by its Latin name, *aurum*, "shining dawn." Mythically, it evokes Midas, the Golden Fleece, and the Rosicrucian "Cross of Gold."

Golden Chain of Homer
An influential 1723 text by Anton Kirchweger, the *Golden Chain of Homer* presents an ordering of the universe from chaos to the quintessence.

gradation
The classification of phases of purification or transformation of a substance by degrees or grades.

grades of fire
The four grades of fire given in seventeenth-century alchemy: *aerial, cineris, igne aperto*, and *balneum Mariae*, relating to the airy, earthly, fiery, and watery characteristics, respectively.

grain
A measure of weight derived from a single grain of barley, being one seven-thousandth of a pound, or approximately sixty-five milligrams. It is still used commercially in the measure of gunpowder, fencing and archery equipment, and pharmaceutical prescriptions.

granulation
The process of rendering a substance to powder, usually through crushing, detonation, or dessication.

Great Work
Specifically, the preparation of the philosopher's stone; a general term for alchemy itself; a broader term for the restoration or union of the alchemist with the Divine.

green lion
A symbol used frequently in alchemical texts to illustrate vitriol (sulfuric acid). It also represents raw, natural energy.

grinding
The reduction (granulation) of a substance by means of friction and pressure, most commonly with a mortar and pestle.

gum
Unless otherwise specified, gum arabic, hardened acacia sap. This is widely used commercially in paints, food, and fireworks.

gypsum
A very soft mineral used in alchemical filtration and as a coagulant. From Hebrew *gephes*, "plaster."

H

Be reserved and silent; Work in a remote, private home; Choose your working hours with prudence; Be patient, watchful, and tenacious; Work on a fixed plan; Use only glass or glazed earthenware crucibles; be rich enough to afford your experiments, or marry a rich wife; Have nothing to do with princes and nobles.
—"The Eight Rules of Albertus Magnus," *Alchemiae Basica*

hartshorn
The horn of the male red deer. Hartshorn oil is distilled from the deer's bones, and salt of hartshorn is either ammonium chloride or ammonium carbonate.

heliotrope
Bloodstone, a form of chalcedony with bright red inclusions of iron oxide (*see* hematite); also a color, the vivid lavender of the heliotrope flower, so named for its growth (tropism) toward the sun (Helios).

hematite
The crystalline form of iron oxide. From Greek *aima*, "blood," for its reddish color.

Hermes Trismegistus
A syncretic combination of the Egyptian god-form Djehuti (Thoth) and Hermes, the Greek god of communication, commerce, and theft, identified with Mercury, the god of learning, language, astrology, and magic. Numerous pseudepigraphic writings attributed to Hermes Trismegis-

tus survive, including the *Corpus Hermeticum*, *Poimandres* (The Divine Pymander), and *Asclepius*. Hermetic monism (all coming from the One) as a philosophy can be said to be the underpinnings of the alchemists' worldview and the work of the reintegration. Perhaps the most famous of the Hermetic texts is the "Emerald Tablet"; one can easily determine its influence on later alchemy:

This is true and remote from all cover of falsehood.

Whatever is below is similar to that which is above. Through this the marvels of the work of one thing are procured and perfected.

Also, as all things are made from one, by the consideration of one, so all things were made from this one, by conjunction.

The father of it is the sun, the mother the moon.

The wind bore it in the womb. Its nurse is the earth, the mother of all perfection.

Its power is perfected.

If it is turned into earth,

Separate the earth from the fire, the subtle and thin from the crude and coarse, prudently, with modesty and wisdom.

This ascends from the earth into the sky and again descends from the sky to the earth, and receives the power and efficacy of things above and of things below.

By this means you will acquire the glory of the whole world, and so you will drive away all shadows and blindness.

For this by its fortitude snatches the palm from all other fortitude and power. For it is able to penetrate and subdue everything subtle and everything crude and hard.

By this means the world was founded

And hence the marvelous conjunctions of it and admirable effects, since this is the way by which these marvels may be brought about.

And because of this they have called me Hermes Trismegistus since I have the three parts of the wisdom and Philosophy of the whole universe.

My speech is finished which I have spoken concerning the Solar Work.[6]

hippocratic wine
Wine mixed with cinnamon, ginger, and other spices; also known as hypocras or piment (Late Middle English). Named for the Greek Hippocrates, the fifth-century BCE father of Western medicine.

Hollandus, Johann Isaac (dates unknown)
Author of several texts on the philosopher's stone, included in the *Theatrum Chemicum* of 1602.

My child shall know, that the Stone called the Philosopher's Stone, comes out of Saturn. And therefore when it is perfected, it makes projection as well in mans' Body from all Diseases, which may assault them either within or without, be they what they will, or called by what name soever, as also in the imperfect Metals.[7]

homunculus
Any depiction, production, or vegetative specimen resembling a small human figure. Some alchemists concerned their research with the creation of a living homunculus, such as the Jewish mythical golem, a crudely formed figure animated by way of an inscription (*amet*, Hebrew, "truth") on the creature's forehead. Literally, Latin, "little man."

humectation
The process of absorbing moisture; the opposite of dessication. From Latin *umere*, "moisten."

hydrometer
A device consisting of a vial and a floating gauge, used to determine the relative density (specific gravity) of liquids compared to the density of water, based on Archimedes's principle of displacement. Literally, Greek, "water measure."

Hypatia of Alexandria (c. 360–415)

A great Neoplatonic scholar, mathematician, logician, and alchemist, Hypatia was one of the most brilliant thinkers of classical antiquity. She is the inventor of the hydrometer and an apparatus for distillation. Hypatia was murdered by a Christian mob in 415.

Alchemy may be compared to the man who told his sons that he had left them gold buried somewhere in his vineyard; where they, by digging found no gold, but by turning up the mould about the roots of the vines, procured a plentiful vintage. So the search and endeavours to make gold have brought many useful inventions and instructive experiments to light.
—Francis Bacon,
De Augmentis Scientiarum, 1623

ignis fortis
Latin, "strong fire."

ignis lentus
Latin, "slow heat."

ignition
The process of initiating combustion so that a substance is consumed. From Latin *ignire*, "to set fire."

imbibation
The process of continually adding a substance to a vessel. From Latin *bibere*, "to drink."

impastation
The process of congealing into pitch or tar; or the product of a mixture with the consistency of paste. From Latin *pasta*, "paste."

impregnation
The state immediately after copulation. From Latin *pre-*, before and *gnasci*, to be born.

inceration

The process of softening to a waxy consistency; or the process of sealing or covering with wax. From Latin *cera*, "wax."

incineration
Reduction to ash. From Latin *cineris*, "ashes." *See* cineration.

incorporation
The process of mixing substances into a compound mass. From Latin *corpus*, "body."

infusion
A suspension in liquid, such as a tea or tincture. From Latin *fusus*, a form of the verb *fundere*, "to melt."

ingression
The irreversible conjunction of substances. From Latin *ingredi*, "to enter."

inhumation
The process of burying. From Latin *humus*, "soil."

iosis
The third or purple phase of the Great Work; cf. coagulation. From Greek *ion*, "violet."

iron
One of the seven alchemical metals, associated with the planet Mars. Iron has been worked in various civilizations for almost four thousand years. Strengthening of iron was a common commercial pursuit of alchemical inquiry.

> *Then, said Nature, without mistake, my son in law, you needs must learn to know the seven planets, of which Mercury is the principal, their powers, their infirmities, their changeable qualities. Tis needful afterward to learn whence Sulphur, Salt and Oil do come. Wherefore we put you in mind of what you will still have occasion for. Sulphur is mighty necessary: so will it give you profit or much ado to make it. Without Salt you shall bring to pass nothing useful for your work. From Oil you have a great mystery. You will make without it nothing sweet-scented. This you ought to remember well, if you would arrive at our Work.*
> —Allegory of John of the Fountain, 1590

Jung, Carl Gustav (1875–1961)

The brilliant Swiss psychiatrist and the father of analytical psychiatry, Jung was heavily influenced by medieval alchemy. He stressed the integration of symbolism, myth, and metaphor into a healthy worldview and held that humanity was by its nature religious. In questing for a template or precedent for his ideas, he settled upon alchemy as a model:

> First I had to find evidence for the historical prefiguration of my own inner experiences. That is to say, I had to ask myself, "Where have my particular premises already occurred in history?" If I had not succeeded in finding such evidence, I would never have been able to substantiate my ideas. Therefore, my encounter with alchemy was decisive for me, as it provided me with the historical basis which I hitherto lacked. . . .

Only after I had familiarized myself with alchemy did I realize that the unconscious is a process, and that the psyche is transformed or developed by the relationship of the ego to the contents of the unconscious. In individual cases that transformation can be read from dreams and fantasies. In collective life it has left its deposit principally in the various religious systems and their changing symbols. Through the study of these collective transformation processes and through understanding of alchemical symbolism I arrived at the central concept of my psychology: the process of individuation.[8]

Jung employed alchemical terminology in his own understanding of the development of the psyche:

The situation is now gradually illuminated as is a dark night by the rising moon. The illumination comes to a certain extent from the unconscious, since it is mainly dreams that put us on the track of enlightenment. This dawning light corresponds to the *albedo*, the moonlight which in the opinion of some alchemists heralds the rising sun. The growing redness (*rubedo*) which now follows denotes an increase of warmth and light coming from the sun, consciousness. This corresponds to the increasing participation of consciousness, which now begins to react emotionally to the contents produced by the unconscious. At first the process of integration is a "fiery" conflict, but gradually it leads over to the "melting" or synthesis of the opposites. The alchemists termed this the *rubedo*, in which the marriage of the red man and the white woman, Sol and Luna, is consummated. Although the opposites flee from one another they strive for balance, since a state of conflict is too inimical to life to be endured indefinitely. They do this by wearing each other out: the one eats the other, like the two dragons or the other ravenous beasts of alchemical symbolism.[9]

A collection of Jung's writings on Chinese and Western alchemy is published under the title *Alchemical Studies*.

♃ Jupiter

Tin. As a planet, Jupiter is associated with expansion, fulfillment, and the qabalistic sphere of *chesed* (mercy). In Roman religion, Jupiter is the son of Saturn and father of Mercury.

Nature is not visible, though it operates visibly; for it is simply a volatile spirit, fulfilling its office in bodies, and animated by the universal spirit—the divine breath, the central and universal fire, which vivifies all things that exist.
—*Alchemical Catechism* ascribed to Paracelsus

Kelley, Edward (1555–97)
An enigmatic figure whose career began in partnership with John Dee, Kelley had in his possession an alchemical text ("the Book of Dunstan") and a red powder, presumably of the philosopher's stone. Dee valued Kelley for his intuitive ability and his visions of the angelic realm, although it was alchemy that most engaged Kelley. He was imprisoned for his failure to produce gold for his patrons, and legend has it he died attempting his escape.

kelp
Seaweed, either dessicated or fermented, used as a source of iodine.

king's yellow
A mixture of white arsenic and orpiment.

Kircher, Athanasius (1601–80)
A German polymath, Jesuit priest, and orientalist, Kircher is the author of the alchemical geology text *Mundus Subterraneus* (1665). His insatiable curiosity led him to develop an early microscope (which enabled him to discover the mechanism of infection in both malaria and plague),

a number of clocks, and the megaphone. He was the first Westerner to discern the connection between Egyptian hieroglyphics and the Coptic language, making him the father of modern Egyptology.

Kirchweger, Anton Joseph (dates unknown)
The author of the *Golden Chain of Homer*, a brief treatise on the alchemical nature and order of material.

Kunckel, Johann (c. 1630–1703)
The son of a court alchemist, Kunckel himself became attached to various European courts. He deciphered Hennig Brand's formula for the production of phosphorous and developed the idea of the alkahest. He reportedly achieved at least three transformations of base metals to gold and made several observations regarding the nature of putrefaction and fermentation.

L

For Hermes said of this Science: Alchemy is a Corporal Science simply composed of one and by one, naturally conjoining things more precious, by knowledge and effect, and converting them by a natural commixtion into a better kind. A certain other said: Alchemy is a Science, teaching how to transform any kind of metal into another: and that by a proper medicine, as it appeared by many Philosophers' Books. Alchemy therefore is a science teaching how to make and compound a certain medicine, which is called Elixir, the which when it is cast upon metals or imperfect bodies, does fully perfect them in the very projection.
—Roger Bacon,
The Mirror of Alchemy, 1597

☿ lapis
In alchemical texts, usually the philosopher's stone. Literally, Latin, "stone."

lavabo
A vessel for washing one's hands in preparation for the Great Work, as a means of ritual purification. Literally, Latin, "I shall wash," derived from the Twenty-Sixth Psalm.

leach
To drain away a substance from ash or soil by way of saturation and percolation. From Old English *leccan*, "to water."

lead
A heavy, dull, poisonous, and malleable metal, lead is associated alchemically with the planet Saturn and thereby with death, limit, *nigredo*, and putrefaction. Given its chemical proximity to gold, it was the most common candidate for transformation.

Leukosis
Albedo, the whitening phase of the Great Work. From Greek *leukos*, "white."

lime
Any mineral consisting primarily of calcium; chalk. Lime was used alchemically as a filter or to reduce acidity in a mixture. Not to be confused with quicklime, calcium oxide. From Latin *limus*, "mud," referring here to mortar.

liquifaction
An aspect of putrefaction, the alchemical process of converting a substance to liquid. From Latin *liquere*, "to flow."

liquor hepatis
A solution of sulfur, lime, and sal ammoniac, which when distilled did not precipitate. This suggested a primal, spiritual nature and was thought to be (or mimic) the presence of the soul, suggesting that the liquor itself was somehow the ethereal made manifest. When incerated, it was known as the balsam of the alchemists. From Latin *liquere*, "to flow," and Greek *hepatos*, "liver."

litharge
Residue from any operation involving metals. From Greek *lithos*, "stone."

lixiviation
The process of separating insoluble substances from the soluble in a compound by means of a solvent; the washing away of soluble material. From Latin *lix*, "lye."

lumen naturae
A kind of moving spirit immanent in matter, an interior light or essence within the natural world posited by some alchemists to exist in addition to a divine light that was to be encountered. Literally, Latin, "light of nature."

lute
A sealant paste made of flour, lime, water, and albumen, used for stopping up bottles. From Arabic *al-oud* or *ud*, "wood."

luting
Sealing a vessel with lute paste. From Arabic *al-oud* or *ud*, "wood."

lutum sapientiae
The bottling of wisdom; a euphemism for alchemy itself. Some alchemical texts were sealed with this glyph.

lye
A highly corrosive alkali salt; caustic soda or potash, derived from the ashes of hardwoods. Lye generates sudden and high temperatures when mixed with water and an explosive hydrogen gas when mixed with aluminum; it can corrode glass.

Let it be for you a great and high mystery in the light of nature that a thing can completely lose and forfeit its form and shape, only to arise subsequently out of nothing and become something whose potency and virtue is far nobler than what it was in the beginning.

—Paracelsus

magnes
Lodestone; a piece of magnetized magnetite (an iron oxide). From Magnesia, the name of a region in central Greece where magnetic ore was first mined.

magnesia
Transformative energy, considered to be a kind of universal principle. The term was derived from a fascination with the power of magnetism and from Magnesia, the name of a region in central Greece where magnetic ore was first mined.

Magnum Opus
Latin, "Great Work."

Maier, Michael (1568–1622)
An alchemist and physician and the author of *Atalanta fugiens* (an illustrated book on alchemical symbolism), Maier was born in Germany but traveled to Basel for his doctorate in medicine. He entered Prague in service to Rudolf II Habsburg and later traveled to England to the court of James I. A Rosicrucian, he authored a text on Hermes Trismegistus and was a reported influence on Sir Isaac Newton.

marcasite
Iron sulfide; a light, brittle crystal, present in both silver-white (*argentea*) and yellow (*aurea*) forms. From Persian *markas*.

Maria Prophetissima (dates unknown)
"Mary the Jewess," a famed alchemist of the early centuries, sometimes identified with Mary Magdalene. She is mentioned by Zosimos and was the inventor of the bain-marie and the still as well as the discoverer of hydrochloric acid. She is quoted as teaching alchemy as the union of opposites: "Join the male and the female, and you shall find that which is sought." She is also credited with the alchemical Axiom of Maria: "One becomes two, two becomes three, and out of the three comes the one, as the fourth," which Jung took as an illustration of the process of individuation.[10]

Mars
Iron. As a planet, Mars is associated with combativeness and with the qabalistic sphere of *geburah* (severity). Mars is the Roman god of war.

materia
The created world (not "matter" as we understand it).

matrass
A bolthead flask with a round base and an elongated neck. The shape is reminiscent of a drawn crossbow (French *materas*), hence the name.

maturation
The progress in an operation.

melanosis
The *nigredo* (blackening phase) of the Great Work. From Greek *melas*, "black."

melting
Fusion; the phase change from solid to liquid. From Proto-Indo-European *meld*, "to soften."

menstruum
A theoretical alkahest that both dissolves and coagulates simultaneously. It was believed that female menstruation was related to fertility in the womb, where sperm and ova were at once dissolved individually and gelled into life. From Latin *mensis*, "month," by way of Greek *mene*, "moon."

Mercury
One of the seven metals of alchemy. Mercury is a toxic, heavy, silver element that is liquid at room temperature. It has been used in extracting silver from ore since the mid-sixteenth century and is obtained by heating cinnabar in a vessel and condensing the vapor.

Mercury is also one of the three principles of alchemy (salt, mercury, and sulfur), acting as a kind of medium or aetheric conduit between salt (*materia*) and sulfur (spirit). It relates to the powers of the mind and language and to the qabalistic sphere of *hod* (glory). Mercury is the Roman god equivalent to the Greek Hermes.

metals
The seven metals of alchemy: copper, gold, iron, lead, mercury, silver, and tin, usually symbolized by their associated planets (Venus, the sun, Mars, Saturn, Mercury, the moon, and Jupiter, respectively).

mineralia
Any combination of the four elements that is devoid of spirit; dead matter. From Latin *minera*, "mine."

minium
An alternate name for cinnabar; a later term for lead oxide. Minium was named for the Minius River in Spain.

mixture
The reversible composition of two or more substances.

moon
Silver. As a celestial body, the moon represents the feminine aspect, dreams, intuition, and the qabalistic sphere of *yesod* (foundation).

mortification
Necrosis, or tissue death. From Latin *mortis*, "death."

mountain blue
Azurite (*lapis armenius*), an ore of copper.

multiplication
The increase in mass of the philosopher's stone.

mutability
The changeability of a substance. From Latin *mutare*, "to change."

Mylius, Johann Daniel (1583–1642)
A lutenist, composer, and alchemist, and the author of *Opus Medico-Chymicum,* published in 1618.

> *Whoever wishes to possess this secret Golden Fleece, which has virtue to transmute metals into gold, should know that our Stone is nothing but gold digested to the highest degree of purity and subtle fixation to which it can be brought by Nature and the highest effort of Art; and this gold thus perfected is called "our gold," no longer vulgar, and is the ultimate goal of Nature.*
> —Eugenius Philalethes,
> An Open Entrance to the Closed Palace of the King, 1645

Natron
Naturally occurring sodium carbonate, a dessiccation agent used by the ancient Egyptians in mummification. From ancient Egyptian *netjeri*, "of the gods."

Newton, Sir Isaac (1643–1727)
A physicist, mathematician, philosopher, astronomer, and alchemist, Newton was one of the most influential thinkers of the Western world and a champion of the scientific revolution. His 1687 *Principia* stands as one of the greatest works in all of science. Newton's Hermetic studies and inquiry into the alchemical aether led him to develop his theory of gravity.

nickel
A silver-white elemental metal, mistaken in ancient times to be a copper ore.

nigredo
The blackening phase of the Great Work, related to putrefaction. From Greek *nekros*, "corpse."

niter
Saltpeter (potassium nitrate), an ingredient in gunpowder. From Hebrew *nether*, "niter," likely from the Egyptian *ntr*.

In the year of our Lord 1382, April 25, at five in the afternoon, this mercury I truly transmuted into almost as much gold, much better, indeed, than common gold, more soft also, and more pliable.
—Nicolas Flamel

ochre
A natural mineral pigment, most commonly iron oxide, used since antiquity. From Greek *ochre*, "brownish yellow."

oil
Any viscous, liquid fat insoluble in water; when not specified in alchemical texts likely refers to olive oil. From Greek *elaia*, "olive."

oil of vitriol
Sulfuric acid.

operation
Any alchemical undertaking. From Latin *opera*, "work."

opus minor
Spagyria, plant alchemy. Literally, Latin, "lesser work."

ormolu
The process of gilding bronze by way of adding granulated gold to a mercury solution, then cooking it onto the bronze object to drive off the

mercury. This results in extremely toxic mercury fumes. From Latin *or*, "gold," and *moulu*, "grounds" or "powder."

orpiment
Arsenic trisulfide; arsenic oxides; a yellow mineral used as dye. From Latin *aurum*, "gold," and *pigmentum*, "pigment."

ouroboros
The alchemical glyph of a serpent or dragon eating its own tail, denoting infinity and specifically the cyclical nature of material existence. In his *Collected Works* Jung writes,

> The Ouroboros has been said to have a meaning of infinity or wholeness. In the age-old image of the Ouroboros lies the thought of devouring oneself and turning oneself into a circulatory process, for it was clear to the more astute alchemists that the *prima materia* of the art was man himself. The Ouroboros is a dramatic symbol for the integration and assimilation of the opposite, i.e. of the shadow. This 'feed-back' process is at the same time a symbol of immortality, since it is said of the Ouroboros that he slays himself and brings himself to life, fertilizes himself and gives birth to himself. He symbolizes the One, who proceeds from the clash of opposites, and he therefore constitutes the secret of the *prima materia*.[11]

> *The Main Purpose and Finis Ultimus of the Hermetic Art is not to produce Gold, as it is the belief of the ill-informed lovers of Gold, but instead the study of God's beautiful Miracles.*
> —*Axiomata*, 1736

Pantheo, Giovanni (dates unknown)

The Venetian author of the *Voarchadumia Contra Alchimiam*, published in 1530. Pantheo's text was an influence on the writings of John Dee, who often mentioned the *Voarchadumia Contra Alchimiam* in his diaries.

Paracelsus (1493–1541)

The Swiss-born Theophrastus Phillippus Aureolus Bombastus von Hohenheim, a Rennaissance polymath and alchemist. He is best known by the (more memorable) name Paracelsus, or "equal of Celsus" (the first-century Roman encyclopedist). His contributions to chemistry, biology, botany, and of course medicine define him as one of the greatest contributors to the later Scientific Revolution. It was Paracelsus who gave us the three principles of salt, mercury, and sulfur. He was also the first scientist to allude to the existence of the unconscious. The hero of the Rosicrucian narratives, Christian Rosenkreutz, is possibly based on Paracelsus. His alchemical works include *Concerning the Tincture of the Philosophers*, *The Treasure of Treasures*, and *The Alchemical Catechism*, among others.

☿ pars cum parte

Electrum, an alloy of gold with no less than twenty percent silver. Martinus Rulandus, in his *Lexicon Alchemiae* (1612), describes it as "a composition of equal parts gold and silver."[12] From Latin, suggesting "side by side."

part
A measure or component volume relative to a whole. "One part x to two parts y" would mean that x composes one-third of the total. From Latin *partiri*, "to divide" or "to share."

paste
Any soft, moist substance, usually derived from the mixture of dry powders in oil, water, egg whites, or another liquid. From Greek *pastos*, "sprinkle."

pater
The masculine element; sulfur. Literally, Latin, "father."

patina
Surface sheen, usually of copper sulfite. Literally, Latin, "shallow pan."

peacock's tail
A symbol for iridescence; the shimmering surface colors observed after the *leukosis* (*albedo*, or whitening phase) but before the *xanthosis* (*citrinas*, or yellowing phase).

pearl
A lustrous, precious gem of calcium carbonate, which forms inside the shell of a bivalve.

pelican
A distillation vessel with two side-tubes that return condensed vapors back into the mass. Its profile resembles a pelican impaling its breast with its beak, mistakenly thought to be an act of nourishing its young with its own blood. This was an esoteric symbol for the sacrifice of Christ, and to the alchemist, personally, a symbol of self-sacrifice for the Work.

phases
Melanosis, xanthosis, leukosis, iosis: blackening, yellowing, whitening, and purpling/reddening, respectively. During the Great Work of alchemical operation, each phase includes specific processes: *melanosis* includes calcinations and putrefaction; *xanthosis* includes fermentation; *leukosis* includes distillation; and *iosis* includes coagulation. So the object of alchemical work is rendered and broken down, processed and fermented, distilled into purity, and solidified into result. The phases are also referred to as the *nigredo, citrinas, albedo,* and *rubedo.* From Greek *phainein,* "to appear."

Philalethes, Eugenius (dates unknown)
Pseudonymous Rosicrucian alchemist and author of *A Short Enquiry into the Hermetic Art* (1714) and other works, such as *Anima Magica Abscondita, Euphrates,* and *Aula Lucis.* Most alchemical historians tend to agree that Philalethes was the pen name for Thomas Vaughan.

philosopher's stone
Lapis philosophorum, the legendary holy grail of alchemy, a substance that transmutes base substances into gold and renders one immortal. It stands at the center of alchemical symbolism and represents perfection, attainment, enlightenment, complete understanding (*gnosis*), and reunion with the Divine. Paracelsus proposed that the stone represented a kind of undiscovered ur-element from which all other elements derived. It is the subject of the Hermetic "Emerald Tablet," first mentioned in the tenth-century *Kitab sirr al-asrar* (The Secret of Secrets). In 1652, William Gratacolle writes,

> Also the Stone is called Chaos, a Dragon, a Serpent, a Toad, the green Lion, the quintessence, our stone Lunare, Camelion, most vild black, blacker than black, Virgins milke, radicall humidity, unctuous moysture, liquor, seminall, Salarmoniack, our Sulfur, Naptha, a soule, a Basilisk, Adder, Secundine, Bloud, Sperme, Metteline, haire, urine, poyson, water of wise men, minerall water, Antimony, stinking menstrues, Lead of Philosophers, Sal, Mercury, our Gold, Lune, a bird, our ghost, dun Salt, Alome of Spaine, attrement, dew of heavenly grace, the stinking spirit, Borax, Mercury cor-

porall, wine, dry water, water metelline, an Egge, old water, perminent, Hermes bird, the lesse world, Campher, water of life, Auripigment, a body cynaper, and almost with other infinite names of pleasure.[13]

phlogiston
A theoretical substance first proposed by Johann Joachim Becher in 1667, combining the ideas of elemental air and fire; inherent combustibility. According to phlogiston theory, once the phlogiston is removed from a substance, the substance is revealed in its true form. From Greek *phlogizein*, "to burn."

phoenix
A symbol of the completion of the Great Work; transformation, renewal, resurrection into eternal life. The phoenix is also used to denote alchemy in general. Possibly from Greek *phoinos*, "blood red."

phosphorus
A nonmetallic chemical element essential for life in all cells, discovered by Hennig Brand in 1669. Literally, Greek, "light bearer" (the Morning Star, Venus).

Picatrix
An eleventh-century Hermetic grimoire (magical book or "grammar") of Arabic origin, consisting of prayers to planetary forces and talismans for controlling them; also, the author of that manuscript. Etymology uncertain.

planets
The "seven stars"; those celestial objects observable by naked-eye astronomy: the moon, the sun, Mercury, Venus, Mars, Jupiter, and Saturn. When we encounter these in alchemical texts (rather than astrological texts) they most commonly refer to the alchemical metals with which they're associated—silver, gold, mercury, copper, iron, tin, and lead.

platinum
A malleable, ductile, grey-white noble metal first identified in 1741. Its symbol is the union of silver and gold.

Pliny (the Elder; 23–79)
A Roman natural historian and military commander, and the author of *Naturalis Historia*, an early encyclopedia.

plumbago
A flowering plant (Plumbaginaceae) thought by Pliny to serve as a cure for lead poisoning. Also called leadwort. From Latin *plumbum*, "lead."

potable gold
A bright red suspension of gold particles in water, also called colloidal gold, used in staining glass and as a medicine.

potash
Any mineral salt containing high levels of potassium. Potash is extracted from the boiled ashes of vegetable waste and used in soap manufacture, textile bleaching, and glassmaking. From Dutch *potaschen*, "ashes in a pot."

potassium
An alkali-metallic element that reacts with water, generating heat. Potassium was unknown as an element by the classical alchemists; its elemental nature was not discovered until the early nineteenth century.

pound
A unit of weight, derived from the Roman *libra* (hence the abbreviation "lb."). From Latin *pondus*, "weight."

powder
Any material of a finely crushed consistency. From Latin *pulvis*, "dust."

precipitation
The formation of a solid in a solution and usually the descension of the precipitate (the solid; the liquid is called the supernate).

From Latin *pre-* "forth," and *caput*, "head," used in the sense "rushing headlong" or "falling suddenly."

pregnation
The onset of putrefaction. From Latin *pre-*, "before," and *gnasci*, "to be born."

preparation
Any substance composed deliberately, more commonly referring to the object of the action rather than the action itself. From Latin *parere*, "to produce" or "to give birth."

priapus tauri
A medicine for impotence. Literally, Latin, "bull penis."

prima materia
The substance of origins; the primeval. Philosophically this can refer to base matter or to the Platonic ideal of matter, which contains an "image" of its own refinement. Literally, Latin, "first matter."

principia
The principles, or *tria prima*, namely salt (substantiation, fixation), sulfur (reaction), and mercury (transformation). These can be understood as body, soul, and spirit, respectively. Literally, Latin, "first."

Principia Chymica
A 1701 text by Johann Christoph Sommerhoff, in which he identifies the alchemical principles as salt, sulfur, mercury, spirit, and earth. The majority of the symbols used in this present volume are extracted from Sommerhoff's work.

processes
The seven alchemical processes, most commonly identified as calcination, sublimation, solution, putrefaction, distillation, coagulation, and tincturation. From Latin *cede*, "to yield" or "to give up."

projection
The rapid addition of substances to a preparation. Literally, a "tossing in" of ingredients, from Latin *jacere*, "to throw."

prolectation
The extraction by thinning of the subtle part from the gross in order to create a lighter substance. Related to Latin *legere*, "to read" or "to discern."

pugillum
A unit of measure, that which can be held between the thumb and the first two fingers. Literally, Latin, "fistful."

pulverization
The reduction of a substance to small particles, usually by grinding or crushing. From Latin *pulvis*, "dust."

purgation
The process of purifying, either through distillation or filtration. From Latin *purgare*, "to cleanse" or "to purify."

purification
The removal of contaminants from a substance. *See* purgation.

putrefaction
Part of fermentation generally regarded as one of the core alchemical processes. Putrefaction speaks to the decay (rotting) of organic matter. From Latin *putrere*, "to rot."

pyrite
"Fool's gold," or iron disulfide, a reflective yellow mineral forming cubic crystals. From Greek *pur*, "fire."

pyrophorus

A substance that spontaneously combusts when in contact with oxygen, made of alum, carbon, flour, and sugar. Literally, Greek, "to bear fire."

Pythagoras (570–490 BCE)

A Greek philosopher, scientist, and mathematician and the developer of a system of ethics and inquiry. As a religious leader, Pythagoras syncretized Egyptian religion and the Orphic cults. He served as a kind of patron saint to the alchemists as he sought to penetrate the mysteries of an inherently divine universe through science. About Pythagoras, Aristotle writes:

> The so-called Pythagoreans, who were the first to take up mathematics, not only advanced this subject, but saturated with it, they fancied that the principles of mathematics were the principles of all things.[14]

The whole secret consists of the measure and government of this thing; that is called her receptacle, and at times, also her fire. This thing is invisible. Due to a certain reverence, it is removed from the eye, and if it comes into the approximate vicinity of somebody, it will retreat to the side in her natural manner, because it is the Secret of Nature.
—Eugenius Philalethes,
Aula Lucis, 1651

quadratum
One-fourth part. From Latin *quadrus*, "square," from *quattuor*, "four."

qualities
Generally, the Platonic qualities of hot, dry, wet, and cold.

 ### quicklime
Calcium oxide, a caustic white chemical compound that reacts violently with water.

quicksilver
Metallic mercury in its liquid state.

quintessence
The "fifth essence" after earth, air, fire, and water, a hypothetical substance or process thought to suspend and vivify all other elements; spirit. From Latin *quinta essencia*, "fifth essence."

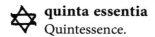 **quinta essentia**
Quintessence.

R

Whosoever, has no knowledge of the rightful beginning, will never find the desired ending; and whosoever, does not know, what he seeks, does also not know, what he will find.
—Axiomata, 1736

rarefaction
Expansion (the opposite of compression). From Latin *rarefacere*, "to make rare."

realgar
"Ruby of arsenic," an orange-red granular mineral; arsenic sulfide. From Arabic *rahj al-gar*, "cave dust."

rebis
The alchemical hermaphrodite, usually depicted as a human with two heads, male and female, and corresponding genitalia. It illustrates the conjunction of sulfur and mercury; the archetype of the Divine Child, the product of the sacred marriage of Sol and Luna, of man and woman; the restoration of all things to their complement, and thereby fruition. An amalgam of the Latin *res bina*, "double thing." (The term *hermaphrodite* itself is the conjunction of the names of the Greek gods Hermes and Aphrodite.)

rectifaction
The purification of a substance through repeated distillation. From Latin *rectificare*, "to make right."

red art
Sulfur; occasionally the *rubedo*.

red lead
Minium (lead oxide), a toxic red pigment used in paint and glass. Named from the Minius River in Spain.

red lion
The *rubedo* (red and final phase) of the Great Work. In this depiction, it is the male principle and represents the phallus in sexual alchemy.

red orpiment
See realgar.

red phase
See rubedo.

reiteration
A cycling through of processes; repetition. From Latin *iterum*, "again."

resolution
The repelling separation of substances in a mixture, such as oil and water. From Latin *solutio*, "unfastening."

restinction
The process of quenching a white-hot substance in a liquid. From *tinctura*, "dying" or "immersing in liquid to change color."

retort
A spherical glass vessel with a long, downward-sloping neck, used in distillation. The neck acts as a condenser. In this way a liquid is heated, distilled to vapor, condensed in the neck, and recollected in a receiver. From Latin *retortus*, "to turn back."

reverberating furnace
A furnace with a curved roof, which deflects heat onto the substance being heated without exposing it to the fuel directly.

reverberation
The process of heating in a reverberating furnace. From Latin *reverberare*, "to strike back."

revivification
The process of returning to life; reactivation. From Latin *vivere*, "to live."

Ripley, George (dates unknown)
A fifteenth-century English alchemist and philanthropist and the author of the *Opera Omnia Chemica* and other works, among them the Ripley Scroll. This latter work is an invaluable—and beautifully illustrated—alchemical text detailing the preparation of the elixir of life and arguably the most visually inspired artifact of the genre.

rubedo
The ultimate phase of the Great Work. It follows the corruption and destruction of matter and the purification of the essence of a substance, culminating in the maturation or synthesis, wherein the experience of the material is identical with the ideal. From Latin *rubeus*, "red."

rubification
Progress, in the Great Work, in bringing a substance from the white phase to the red phase. From Latin *rubeus*, "red."

Rulandus, Martinus (the Younger; 1569–1611)
A Bavarian-born physician and alchemist and the author of the *Lexicon Alchemiae* (1612). Many of the terms in Rulandus's *Lexicon* seem to be personal ciphers, not common to alchemical texts of the period.

S

> *Here I give you my key and My seal; one opens,
> one locks: Use both with understanding.*
> —Eugenius Philalethes,
> *Lumen de Lumine*, 1651

sal alkali
Caustic potash; lye. From Latin *sal*, "salt," and Arabic *al-kali*, "ash."

sal amoniac
A soft, rare mineral used as flux in glassworking. Literally, Latin, "salt of Amun" (*see* ammonia).

sal marinum
Latin, "sea salt."

salt
An alchemical principle representing the body, the material world, and the white phase of the Great Work. Salt is also used metaphorically for wisdom, self-knowledge, and wit.

saltpeter
Potassium nitrate leached from manure and wood ash, used in fertilizers and gunpowder. Likely from Latin *sal*, "salt," and *petra*, "stone."

♦♦♦ sand
♦♦♦ A substance used in filtration, and a base material from which to extract certain minerals.

♄ sapphire
A (usually) blue precious gemstone used in ornamentation and for its extreme hardness. From Greek *sappheiros*, "blue stone."

♄ Saturn
Lead. As a planet, Saturn refers to time and mortality and corresponds to the qabalistic sphere of *binah* (understanding). Saturn is the Roman god of the harvest, hence his association with reaping and thereby death.

Schweitzer, Johann Friedrich (1625–1709)
A Dutch physician and alchemist and an associate of Spinoza and doctor to the Prince of Orange. Schweitzer is the author of *Vitulus Aureus* (*The Golden Calf*), published in 1667.

℈ scrupulus
A measure of 20 grains. Literally, Latin, "pebble."

segregation
The process of ordering a compound into separate substances. From Latin *se-*, "apart from," and *grex*, "flock" or "herd."

Sendivogius, Michael (1566–1636)
Michał Sędziwój, a Polish alchemist and philosopher. He discovered the principles of air and the existence of oxygen and purportedly transmuted large quantities of base metal into gold for his patron King Sigismund III Vasa. He is the author of *A New Light on Alchemy*, published in 1605.

♄ separation
Division and isolation or removal of substances within a compound.

sepulcher
The philosophical egg, an earthenware vessel or crucible. From Latin *sepelire*, "to bury."

siccity
Dryness. From Latin *siccus*, "dry."

 silex
Silica; quartz or flint. Literally, Latin, "flint."

silver
A precious grey, shiny metallic element associated with the moon and the female principle.

skeleton
In alchemical illustrations, a symbol for the *albedo* (whitening or purifying phase) of the Great Work.

skillet
A flat-bottomed circular pan. From Latin *scutella*, "serving platter."

skull
A symbol for the *caput mortuum*, or waste by-product of alchemical work. The skull also serves as a *"memento mori,"* a reminder of mortality.

sol niger
Aside from the observable sun of the material universe, also a symbol of the *nigredo* (blackening phase) of the Great Work. Literally, Latin, "black sun."

solution
A liquid mixture comprising a medium (the solvent) and a distributed component (the solute). From Latin *solutio*, "unfastening."

soot
Powdered carbon; the remainder of burning. Related to the word "sit" or "settle."

spagyria
The preparation of herbal medicines, most commonly employing fermentation and distillation. The term was coined by Paracelsus. From Greek *spao*, "to tear," and *ageiro*, "to collect."

spirit
The quintessential nature of a person or thing; a distilled liquor. From Latin *spiritus*, "breath."

spirit of wine
Rectified ethyl alcohol.

stagnation
Cessation of movement, or depletion of energy in a process. From Latin *stagnum*, "pool."

steel
A hard alloy of iron, carbon, and other elements such as nickel.

St. Germain, Comte (dates unknown)
An eighteenth-century polymath, occultist, and alchemist. The Comte St. Germain was an enigmatic figure of renown and notoriety. Throughout his career he claimed various noble parentages from Transylvania, Spain, and Portugal. Giacamo Casanova called him "a celebrated and learned imposter."

stibium
Latin, "antimony."

stone
Aside from the common English use of the word, a "shew stone," or gazing stone, such as a crystal ball used in fortune-telling or remote viewing.

stopping
See luting.

stratification
The process of layering a compound into distinct substances or states. From Latin *stratum*, "spreading out."

subduction
The downward separation of a substance from a compound, through pressure or gravity. From Latin *sub-*, "beneath," and *ducere*, "to lead."

sublimation
The process of condensing a vapor into a solid without undergoing a liquid phase. From Latin *sub-*, "up to," and *limen*, "threshold" or "lintel."

substance
A kind of matter with uniform properties. For example, ice cubes and liquid water are the same material, but as they exhibit different properties they are assumed, in alchemical texts, to be distinct substances. A Latin (*substare*, "to stand under") translation of the Greek *hypostasis*, "underlying reality or essence."

subtilation
The process of separating the fine substances from the base or gross substances in a compound. From Latin *titillare*, "to tickle" or "to coax."

sugar
A highly combustible crystalline form of sucrose, extracted from plants.

sulfur
A bright yellow chemical element. sulfur was used in antiquity as a topical antiseptic. As an alchemical principle, it symbolizes the active or vivifying nature.

sun
Gold. As a celestial object, the sun is considered the foremost of the planets, denoting enlightenment and corresponding to the male aspect and the qabalistic sphere of *tiphereth* (beauty). The Greek Pythagorean philosopher Philolaus posited there were two suns: an ideal, philosophical sun; and a material sun, the black sun or *sol niger*.

swan
In alchemical illustrations, a symbol for the *albedo* (whitening or purifying phase) of the Great Work.

T

Now no Errour follows in the Work,
Burn all with a very strong Fire,
Bring out at length the Blood, the Soul
After the White King: Then thrice imbibe.
(The King being thus known) the Body is the Soul,
And fixt, and permanent, although like Wax;
The Colour is not an Accident; but a Substance
Reigning in all, with the highest Glory.
—Thomas Rawlin,
"An Alchemical Poem," 1611

talc
Magnesium silicate, a lustrous mineral used as a soft polish and in filtration. From Persian *talq*.

tartar
A salt of tartaric acid, a white crystalline substance found in grapes and used as a preservative in wine. Tartaric acid was first isolated by Geber in the early ninth century.

terebinth
Resin, or turpentine, of the terebinth tree (*Pistacia terebinthus*).

terra damnata
See caput mortuum.

test
In alchemical parlance, the equivalent of a logical argument or mathematical proof.

Theatrum Chemicum
An encyclopedic collection of eighty-eight alchemical works over six volumes, published between 1602 and 1661. This massive and significant undertaking made available obscure and previously unpublished alchemical manuscripts from a variety of authors, including Thomas Aquinas, John Dee, Roger Bacon, and Albertus Magnus.

tigillum
Crucible. Literally, Latin, "small wooden beam," referring to the Crucifixion.

tin
A common silvery, flexible metallic element, used for its resistance to corrosion and oxidization and in alloys, such as bronze.

tincturation
The process of making a tincture from a "crude drug" (unrefined plant or mineral compound).

tincture
A solution composed of a substance dissolved in alcohol. From Latin *tinctura*, "dye."

toad
In alchemical illustrations, a symbol for putrefaction.

topaz
A colorless silicate mineral, generally confused with the yellow stone chrysolite.

transudation
The sweating of a substance during distillation. From Latin *sudare*, "to sweat."

trituration
The process of rendering a substance into powder by the application of heat instead of crushing. From Latin *tritura*, "rubbing" or "as though rubbed, ground."

Tubal-Cain
The biblical father of metallurgy, mentioned in Genesis 4:22, "Zillah, on her part, gave birth to Tubalcain, the ancestor of all who forge instruments of bronze and iron" (New American Bible). His name may be etymologically related to Vulcan, the Roman god of the smith.

turquoise
A green-blue semiprecious stone, hydroxyl phosphate of aluminum and copper, used in antiquity as ornament. Likely derived from "Turkey."

tutia
Crude zinc oxide used as a polishing powder. Also "tutty." From Arabic *tutiya*.

He that can destroy bodies without putrefaction, and in the destruction can join Spirit with Spirit by means of heat, possesses the principal secret of natural magic.
—Anonymous

ulcer
A discontinuity in the skin or other membrane. From Latin *ulcer-*.

uncia
Ounce, a unit of mass equal to roughly 28 grams. Literally, Latin, "one-twelfth," i.e., one-twelfth of a pound.

unicorn
In alchemical illustrations, a symbol for the *albedo* (whitening or purifying phase) of the Great Work.

union
The combination of two substances into a single compound. From Latin *unus*, "one."

urine
Usually human, unless otherwise specified (frequently equine), used extensively in alchemy as a source of ammonia and phosphorous.

UXOR ODORIFERA

☽ **uxor odorifera**
Silver in its aspect of the feminine principle. Literally, Latin, "fragrant wife."

> *It is impossible, that any Mortal understands this Art, unless he has been previously enlightened by the Divine Light.*
> —Gerardus Dorneus,
> *Theatrum Chemicum Brittannicum*, 1602

Valentine, Basil (c. 1394–?)
A fifteenth-century German priest and alchemist and the author of *The Twelve Keys of Philosophy*, *The Triumphal Chariot of Antimony*, and *Azoth*.

> It will be necessary first of all to utter, and to acquaint you by a speech, that all things consist of two parts, that is, Natural and Supernatural; what is visible, tangible, and hath form or shape, that is natural; but what is intactible, without form, and spiritual, that is supernatural, and must be apprehended and conceived by Faith.[15]

valerian
A hardy green perennial herb (*Valeriana officinalis*) used since antiquity as a sedative.

van Helmont, Jan Baptist (1579–1644)
A Flemish alchemist, mystic, and physician. He coined the word *gas* (from the Greek *chaos*) as it applies to pneumatic chemistry. He is the author of *Ortus Medicinae* (*The Origin of Medicine*), published in 1648.

vapor
A substance diffused in air, such as steam. From Latin *vapor*.

Vaughan, Thomas (1621–66)
A Welsh philosopher, priest, and alchemist, Vaughan was a noted Rosicrucian, translating that organization's texts into English. He is possibly the same as "Eugenius Philalethes," a nom de plum.

Venus
Copper. As a celestial body, Venus corresponds to fertility and the qabalistic sphere of *netzach* (victory). She is the Roman goddess of love, corresponding to the Greek Aphrodite.

verdigris
Copper patina formed by oxidization. Literally, Old French, "green of Greece."

vermilion
A red pigment of cinnabar. From Latin *vermis*, "worm."

Villa Nova, Arnaldus de
See de Villa Nova, Arnaldus.

vinegar
Acetic acid, derived from fermentation. From Latin *vinum*, "wine," and *acer*, "sour."

viride aes
From Latin *viridis*, "green"; and *aes*, "copper," "brass," or "unrefined ore." *See* verdigris.

vitrification
The process of transmuting a substance into glass by the application of heat. From Latin *vitrum*, "glass."

⊕ vitriol
Sulfuric acid; generally, any kind of trial thought to wash away impurities. From Latin *vitrum*, "glass," for its glassy appearance.

✸ vitrum
Latin, "glass."

Voarchadumia
An alchemical and metallurgical text by the Venetian Giovanni Agostino Pantheo, published in 1530, considered to have been an influence on the work of John Dee.

△ volatile
Capable of evaporating at normal temperatures; explosive. From Latin *volare*, "to fly."

Voynich Manuscript
A handwritten and beautifully illustrated cipher-text composed in the fifteenth century. The Voynich manuscript is an historical enigma: while it does appear to be an herbal, possibly spagyric text, it also seems to have astronomical or astrological correspondences. Its interest to alchemical scholarship is in its possible authorship, purportedly either John Dee, Roger Bacon, or a similar figure.

Out of the true Affection and Charity, we have for all Lovers of Arts, we advise every one, who shall desire to prepare either of these our Elixirs, only to follow our Infallible Rules, being the Compendium of the whole Practice and Theory according to all true Philosophers, and not to mind any other: for some having delivered things by Hearsay, others from Reading, and very few from their own Practice, they may easily be imposed upon and deluded by any Pseudochymist or pretended Adept.
—Aphorismi of Urbigerani, 1690

 water
The third of the four classical elements, water is the font of dream, intuition, emotion, and the subconscious. Alchemically its glyph is the inverted triangle of the womb. Plato identified water with the icosahedron and the qualities of cold and wet. Its magical animal is the undine, or mermaid. In tarot, water is represented by the suit of cups, and the angel of water is Gabriel.

 water of life
Whiskey; see *aqua vitae*.

wax
Beeswax. From Old English *weax*.

white art

Commonly mercury; also occasionally the white phase of the Great Work.

white lead

A common, toxic white pigment used in oil painting.

white queen

In alchemical illustrations, a symbol for the *albedo* (whitening or purifying phase) of the Great Work. *See* white stone.

white stone

In alchemical illustrations, a symbol for the *albedo* (whitening or purifying phase) of the Great Work.

 wine

Fermented fruit juices. From Latin *vinum*, "wine."

 wood ash

The carbon remains of wood after burning, used in filtration and in the extraction of potash.

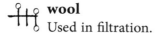 **wool**

Used in filtration.

*The Eagle therefore and the Woman, as likewise the
Dragon with almost all the Severals of the whole Art, are
Understood by these precepts; which by opening the Bosom
of Nature We have perhaps so far Explained and declared to
the Sons of Learning, that so Glory might be given to God.*
—Michael Maier,
Atalanta Fugiens, 1617

xanthosis
Citrinas, or the yellowing phase of the Great Work. Some alchemical texts place *citrinas* between *albedo* and *rubedo,* although it is generally considered an aspect of the red phase. From Greek *xanthos,* "yellow."

To be short—if you attempt this discourse, you do it without my advice, for it is not fitted to your fortunes. There is a white magic this book is enchanted withal: it is an adventure for Knights of the Sun, and the errants of this time may not finish it.
—Eugenius Philalethes,
Aula Lucis, 1651

yellow phase
Not generally considered to be a distinct phase of the Great Work, but rather a precursor to the red or final phase. *See citrinas*, also *xanthosis*.

yolk
The protein-rich, fatty yellow part of a bird's egg. From Old English *geolca*, "yellow."

Z

All real Adepts speak with one voice and if they speak truly, one may, without taking so much trouble, without employing so many vases, without consuming so much charcoal, without ruining one's purse and one's health, one may, I repeat, work in concert with Nature, who, being aided, will lend herself to the desires of the Artist and will freely open to him her treasures. He will learn from her, not how to destroy the bodies which she produces, but how and from what, she composes them, and into what they resolve. She will show him that matter, that chaos from which the Supreme Being has formed the Universe. They will see Nature, as in a mirror, and her reflection will manifest to them the infinite wisdom of the Creator, who directs and guides her, in all her operations, by a simple and unique way which constitutes all the mystery of the Magnum Opus.

—Dom Jacques Pernety,
A Treatise on the Great Art, 1898

zinc
A silvery white metallic element used for its resistance to corrosion.

zodiac
The ring of constellations that mark the progress of the sun as observed from Earth over the course of the year. From Greek *zodiaklos kuklos*, "circle of animals." Alchemists were mindful of the impact of the progressions of the constellations on their Work, out of the Hermetic principle "as above, so below."

ZOSIMOS OF PANOPOLIS (BORN C. 300 AD)

Zosimos of Panopolis (born c. 300 AD)
A Greco-Egyptian Gnostic/Hermetic philosopher, alchemist, and the author of the first-known book on alchemy, composed in the fourth century. Included in this fragmentary surviving text is one of the earliest definitions of alchemy:

> The composition of the waters, and the movement, and the growth, and the removal and restitution of bodily nature, and the splitting off of the spirit from the body, and the fixation of the spirit on the body are not operations with natures alien one from the other, but, like the hard bodies of metals and the moist fluids of plants, are One Thing, of One Nature, acting upon itself. And in this system, of one kind but many colors, is preserved a research of all things, multiple and various, subject to lunar influence and measure of time, which regulates the cessation and growth by which the One Nature transforms itself.[16]

As inspiration to modern or speculative alchemists, Zosimos's greatest surviving contribution came not from a laboratory, but from his dreams. Zosimos documented repeated visions of either himself or a homunculus being repeatedly destroyed by way of torture, only to be transformed into a spiritual being. Jung was inspired by this account and wrote an extraordinary analysis of Zosimos's visions (*The Visions of Zosimos*, 1937).

Notes

Introduction

Epigraph. W. B. Yeats, *Rosa Alchemica*, Project Gutenberg, http://www.gutenberg.org/ebooks/5794 (accessed October, 2006).

1. Richard Rolle de Hampole, 1380, http://www.alchemylab.com/great_work_begins_here.htm (accessed June, 2005).

2. Plato, *Timaeus*, Project Gutenberg, http://www.gutenberg.org/ebooks/1572 (accessed June, 2006).

3. C. G. Jung, *Memories, Dreams, Reflections*, ed. Aniela Jaffé, 1965 (repr. New York: Vintage / Random House, 1989), 201.

4. Albert Camus, in *The Yale Book of Quotations*, ed. Fred R. Shapiro (Cambridge: Yale University Press, 2006), 130.

Text

Epigraph for "A" Entries. Arthur Schopenhauer, http://quotationsbook.com/quote/9543/#axzz1MYGzQcNf (accessed August, 2007).

1. A. E. Waite, quoted in Fred Gettings, *Dictionary of Occult, Hermetic, and Alchemical Sigils* (London: Routledge & Kegan Paul, 1981), 27.

2. Paracelsus, *Hermetic and Alchemical Writings of Paracelsus*, vol. 1, trans. A. E. Waite (Chicago: De Laurence, Scott & Co, 1910), 294.

3. Jakob Boehme, *De Signatura Rerum*, quoted in Robert Ambelain, *Spiritual Alchemy*, trans. Peirs Vaugh (privately printed, 2005), 23.

Epigraph for "B" Entries. Edmund Dickinson, *The Worck of Dickinson*, seventeenth-century writings transcribed from Ferguson Manuscript 91, Glaskow University Library.

Epigraph for "C" Entries. Arnaldus de Villa Nova, *A Chymicall Treatise*, 1611; repr. in *Spiegel der Philosophen* (Mirror of the Philosophers) from Aurei Velleris (Hamburg: bei Christian Liebezeit, in der St. Joh. Kirch, 1708).

Notes

Epigraph for "D" Entries. Thomas Vaughan (a.k.a. Eugenius Philalethes), *Concerning the Hermetic Art,* 1714; quoted in Sean Martin, *Alchemy and Alchemists* (Harpenden, Hertfordshire, UK: Pocket Essentials, 2006), 25.

Epigraph for "E" Entries. A German Sage, *A Tract of Great Price,* 1423; unpublished document in private circulation, transcriber Bauer Lindemann, 2010.

4. Plato, *Timaeus,* trans. Benjamin Jowett, http://classics.mit.edu/Plato/timaeus.html (accessed June, 2009).

Epigraph for "F" Entries. Johann Isaac Hollandus, *A Work of Saturn,* from *Of Natural and Supernatural Things,* transcriber Joshua Ben Arent, London, 1670, http://www.levity.com/alchemy/hollandus_saturn.html (accessed June, 2011).

5. John French, *The Art of Distillation, Or, A Treatise of the Choicest Spagyrical Preparations Performed by Way of Distillation, Being Partly Taken Out of the Most Select Chemical Authors of the Diverse Languages and Partly Out of the Author's Manual Experience together with, The Description of the Chiefest Furnaces and Vessels Used by Ancient and Modern Chemists . . . Printed by Richard Cotes and are to be sold by Thomas Williams at the Bible in Little-Britain without Aldersgate, 1651*; PDF in private circulation, transcriber Bauer Linemann, 2010.

Epigraph for "G" Entries. Aesch-Mezareph, or Purifying Fire, 16th Century, quoted in *Alchemy Journal* 2, no. 4 (July–August 2001): 12.

Epigraph for "H" Entries. "The Eight Rules of Albertus Magnus," in *Alchemiae Basica,* privately-circulated manuscript, compiler Amanda Diane Doerr, date unknown.

6. *The Emerald Tablet of Hermes,* from *Aurelium Occultae Philosophorum,* 1610, trans. Georgio Beato, http://www.blackmaskonline.com (accessed 2002).

7. Johann Isaac Hollandus, *A Work of Saturn (see note for epigraph for "F" entries).*

Epigraph for "I" Entries. Francis Bacon, *De Augmentis Scientiarum,* 1623; quoted in Neil Powell, *Alchemy: The Ancient Science* (London: Aldus Books, 1976), 101.

Epigraph for "J" Entries. Allegory of John of the Fountain, 1590; Sloane MS 3637, British Library, London.

8. Jung, *Memories, Dreams, Reflections,* 200 (see intro., n. 3).

Notes

9. _____. *Analytical Psychology: Its Theory and Practice*, ed. William McGuire, 1926; repr. Bollingen Series 99 (Princeton: Princeton University Press, 1991), 22–23.

   Epigraph for "K" Entries. **Alchemical Catechism** attributed to Paracelsus quoted in *Hermetic and Alchemical Writings of Paracelsus*, trans. A. E. Waite (Sioux Falls, SD: Nu Vision Publications, 2007), 7.

   Epigraph for "L" Entries. Roger Bacon, *The Mirror of Alchemy*, 1597; http://www.levity.com/alchemy/mirror.html (accessed June 2011).

   Epigraph for "M" Entries. Paracelsus quoted in Jay Ramsay, *Alchemy: The Art of Transformation* (London: Thorsons, 1997), 221.

10. Both of Maria Prophetissima's sayings quoted in H. Fröhlich, "On Isobaric Spin Space," dissertation, Department of Theoretical Physics, University of Liverpool, USA, received April, 1963.

    Epigraph for "N" Entries. Thomas Vaughan (a.k.a. Eugenius Philalethes), *An Open Entrance to the Closed Palace of the King*, 1645 (Sequim, WA: Holmes, 1993), 5.

    Epigraph for "O" Entries. Nicolas Flamel, quoted in Mary Anne Atwood, *Suggestive Inquiry into the Hermetic Mystery with a Dissertation on the More Celebrated of the Alchemical Philosophers, Being an Attempt Towards the Recovery of the Ancient Experiment of Nature*, 1850 (repr. Whitefish, MT: Kessinger, 1999), 43.

11. C. G. Jung, *Mysterium Coniunctionis* trans. Gerhard Adler and R. F. C. Hull, *Collected Works of C. G. Jung*, vol. 14; 2nd edition, Bollingen Series 20 (Princeton: Princeton University Press, 1977), para. 513.

    Epigraph for "P" Entries. Axiomata, 1736; quoted in Johannes Helmond, *Alchemy Unveiled*, trans. Gerard Hanswille and Deborah Brumlich (Canada: Merkur, 1996), 118.

12. Martinus Rulandus, *A Lexicon of* Alchemy, 1612; trans. A. E. Waite (London: John M. Watkins, 1893), 189.

13. William Gratacolle, *Names of the Philosopher's Stone*, in *Five Treatises of the Philosopher's Stone*, London, 1652. http://www.levity.com/alchemy/gratacol.html.

14. Aristotle, *Metaphysics*, trans. Richard Hope (Ann Arbor: University of Michigan Press, 1952), 28.

Notes

Epigraph for "Q" Entries. Thomas Vaughan (a.k.a. Eugenius Philalethes), *Aula Lucis, or, The House of Light: a discourse written in the year 1651 by S. N., a modern speculator . . . London: Printed for William Leake and are to be sold at his shop, at the sign of the Crown in Fleet Street, between the two Temple Gates, 1652. 50 pence; (repr. Whitefish, MT: Kessinger, 2010), 6.*

Epigraph for "R" Entries. Axiomata, 1736; quoted in Helmond, *Alchemy Unveiled*, 27 (see note for epigraph for "P" entries).

Epigraph for "S" Entries. Thomas Vaughan (a.k.a. Eugenius Philalethes), *Lumen de Lumine*, London, 1651; quoted in Helmond, *Alchemy Unveiled*, 39 (see note for epigraph for "P" entries).

Epigraph for "T" Entries. Thomas Rawlin, "An Alchemical Poem," 1611, at the end of "A warning to the False Chymists or the Philosophical Alphabet by Thomas Rawlin, folios 14–55"; Sloane MS 3643, British Library, London. See also http://www.alchemywebsite.com/rawlin.html.

Epigraph for "U" Entries. "An Anonymous Treatise upon Magnetical Physic, divided into three parts," 1321, transcriber Adam McLean, 2nd section, Sloane MS 1321, British Library, London.

Epigraph for "V" Entries. Gerardus Dorneus, *Theatrum Chemicum Brittannicum*, 1602, http://www.gutenberg.org/files/26340/26340-h/26340-h.htm (accessed November 2010).

15. Basil Valentine, *Of Natural and Supernatural Things; Also of the first Tincture, Root, and Spirit of Metals and Minerals, how the same are Conceived, Generated, Brought forth, Changed, and Augmented*, Public Domain Books, 2006, Kindle edition.

Epigraph for "W" Entries. Aphorismi Urbigerani, *Or Certain Rules, Clearly demonstrating the Three Infallible Ways of Preparing the Grand Elixir or Circulatum majus of the Philosophers*, London, 1690; http://www.levity.com/alchemy/urbigeri.html (accessed August 2007).

Epigraph for "X" Entries. Michael Maier, *Atalanta Fugiens*, 1617; Sloane MS 3645, British Library, London.

Epigraph for "Y" Entries. Thomas Vaughan (a.k.a. Eugenius Philalethes), *Aula Lucis* (see note for epigraph for "Q" entries).

Notes

Epigraph for "Z" Entries. Dom Jacques Pernety, *A Treatise on the Great Art; A System of Physics according to Hermetic Philosophy and Theory and Practice of the Magisterium*, 1898 (repr. Flaming Sword, 1997), 33.

16. C. G. Jung, "The Visions of Zosimos," from *Alchemical Studies*, trans. R. F. C. Hull, *Collected Works of C.G. Jung*, vol. 13, Bollingen Series 20 (Princeton: Princeton University Press, 1967), 59.

BIBLIOGRAPHY

Albertus Magnus. *The Book of Secrets of Albertus Magnus: Of the Virtues of Herbs, Stones, and Certain Beasts; Also a Book of the Marvels of the World.* Edited by Michael R. Best and Frank H. Brightman. Boston: Red Wheel/Weiser, 2000.

Allen, Paul M., ed. "Secret Symbols of the Rosicrucians of the 16th and 17th Centuries," in *A Christian Rosenkreutz Anthology.* New York: Rudolf Steiner Publications, 1968.

Apuleius. *The Golden Ass,* translated by Robert Graves. Hammondsworth: Penguin Books, 1976.

Berthelot, Marcelin. *Origines de l'alchimie.* Paris, 1885.

Budge, Sir E. A. Wallis. *Egyptian Magic.* London: Routledge & Kegan Paul, 1972.

Churton, Tobias. *The Golden Builders: Alchemists, Rosicrucians, and the First Freemasons.* Lichfield, UK: Signal Publishing, 2002.

Cotnoir, Brian. *The Weiser Concise Guide to Alchemy.* San Francisco: Red Wheel/Weiser Books, 2006.

Doresse, Jean. *The Secret Books of the Egyptian Gnostics.* New York: MFJ Books, 1986.

Eliade, Mircea. *The Forge and the Crucible,* 2nd ed. Chicago: University of Chicago Press, 1956.

Faivre, Antoine. *Accès de l'ésotérisme occidental.* Paris: Gallimard, 1986.

Flamel, Nicolas. *Hieroglyphical Figures.* Translated by Eirenaeus Orandus for Thomas Walsley, at the Eagle and Child in Britans Burusse. London, 1624.

Freedman, David N. *The Anchor Bible Dictionary.* New York: Doubleday, 1992.

Geber. *The Alchemical Works of Geber.* Translated by Richard Russell. New York: Samuel Weiser, 1994.

Bibliography

Gettings, Fred. *Dictionary of Occult, Hermetic, and Alchemical Symbols.* London: Lowe and Brydone Printers, 1981.

Glaser, Christopher. *The Compleat Chymist.* 1677. Richardson, Texas: Restorers of Alchemical Manuscripts, 1983.

Graves, Robert. *The White Goddess.* London: Faber & Faber, 1961.

Hall, Manly Palmer. *The Hermetic Marriage.* Los Angeles: Philosophical Research Society, 1996.

Holland, Isaac. *A Compendium of Writings by Johan Isaaci Hollandus.* Translated from German by RAMS. Richardson, Texas: Restorers of Alchemical Manuscripts, 1981.

Jung, Carl Gustav. *Alchemical Studies.* Princeton, NJ: Princeton University Press, 1967.

Kirschweger, Anton. *The Golden Chain of Homer.* Translated from the German edition (1723) by Sigmund Bacstrom. 1797.

Levi, Eliphas. *La Science Des Esprits—Revelation du dogme secret des kabbalistes.* Paris: Germer Bailliere, 1865.

Levi, Eliphas. *The History of Magic—Including a Clear and Precise Exposition of Its Procedure, Its Rites, and Its Mysteries.* London: Rider, 1913.

Mackey, Albert G. *Encyclopedia of Freemasonry.* Philadelphia: McClure Publishing, 1917.

Mathers, S. L. MacGregor. *The Kabbalah Unveiled: Containing the Books of the Zohar.* London: Arkana, 1926.

McIntosh, Christopher. *The Rose Cross and the Age of Reason: Eighteenth-Century Rosicrucianism in Central Europe and Its Relationship to the Enlightenment.* Leiden, the Netherlands: E. J. Brill, 1992.

McLean, Adam. "Database of Alchemical Books." http://www.levity.com/alchemy/database.html.

Paracelsus. *The Hermetic and Alchemical Writings of Paracelsus.* Edited by Arthur Edward Waite. London: James Elliot & Co., 1894. Reprint, Whitefish, MT: Kessinger, 2005.

Patai, Raphael. *The Jewish Alchemists.* Princeton, NJ: Princeton University Press, 1994.

Bibliography

Philalethes, Eugenius. *An Open Entrance to the Closed Palace of the King*. Richardson, Texas: Restorers of Alchemical Manuscripts, 1981.

Read, John. *Through Alchemy to Chemistry*. London: G. Bell, 1957.

Regardie, Israel. *The Philosopher's Stone*, 2nd ed. St. Paul, Minnesota: Llewellyn Publications, 1970.

Ripley, George. *Liber Secretisimuss*. Richardson, Texas: Restorers of Alchemical Manuscripts, 1982.

Rowling, J. K. *Harry Potter and the Philosopher's Stone*. London: Bloomsbury, 1997.

Ruland, Martin. *A Lexicon of Alchemy or Alchemical Dictionary*. London: J. M. Watkins, 1964.

Sendivogius, Michael. *The New Chemical Light*. 1608. Reprint, Richardson, Texas: Restorers of Alchemical Manuscripts, 1982.

Shah, Sayed Idries. *The Secret Lore of Magic*. London: Frederick Muller, 1957.

Singer, Isidore, ed. *The Jewish Encyclopedia, New Edition*. New York: Funk & Wagnalls, 1925.

Spence, Lewis. *An Encyclopedia of Occultism*. New York: Dodd, Mead, 1920.

The Testament of Nicholas Flamel. London: Printed by J. & E. Hodson, 1806.

Valentine, Basil. *Triumphant Chariot of Antimony with Annotations of Theodore Kirkringius, MD*. Printed for Dorman Newman at the Kings Arms in the Poultry, 1678.

Yarker, John. *Scientific and Religious Mysteries of Antiquity: The Gnosis and Secret Schools of the Middle Ages; Modern Rosicrucianism; and Free and Accepted Masonry*. London: John Hogg, 1872. Reprint, Whitefish, MT: Kessinger, 1997.

Yates, Dame Frances. *The Rosicrucian Enlightenment*. London: Routledge & Kegan Paul, 1972.

Quest Books

encourages open-minded inquiry into
world religions, philosophy, science, and the arts
in order to understand the wisdom of the ages,
respect the unity of all life, and help people explore
individual spiritual self-transformation.

Its publications are generously supported by
The Kern Foundation,
a trust committed to Theosophical education.

Quest Books is the imprint of
the Theosophical Publishing House,
a division of the Theosophical Society in America.
For information about programs, literature,
on-line study, membership benefits, and international centers,
see www.theosophical.org
or call 800-669-1571 or (outside the U.S.) 630-668-1571.

To order books or a complete Quest catalog,
call 800-669-9425 or (outside the U.S.) 630-665-0130.

Related Quest Titles

A Dictionary of Gnosticism, by Andrew Phillip Smith

The Golden Thread: The Ageless Wisdom of the Western Mystery Traditions, by Joscelyn Godwin

Indra's Net: Alchemy and Chaos Theory as Models for Transformation, by Robin Robertson

Hidden Wisdom: A Guide to the Western Inner Traditions, by Richard Smoley and Jay Kinney

To order books or a complete Quest catalog,
call 800-669-9425 or (outside the U.S.) 630-665-0130.